MW00843959

# FUNDAMENTALS OF
# SYSTEMOLOGY

*Thank you for supporting independent publishing
and the Systemology Society.*

**mardukite.com**

MARDUKITE ACADEMY OF SYSTEMOLOGY COLLECTOR'S EDITION

# FUNDAMENTALS OF SYSTEMOLOGY

## A NEW THOUGHT FOR THE 21ST CENTURY

*Mardukite Academy Basic Course
developed by Joshua Free*

THE JOSHUA FREE IMPRINT
JFI PUBLICATIONS

© 2023, JOSHUA FREE

ISBN : 978-1-961509-24-5

A MARDUKITE SYSTEMOLOGY PUBLICATION

Mardukite Research Library Catalogue No. "Liber-S1A" (1A)

*Developed for Mardukite Academy & The Systemology Society*

cum superiorum privilegio veniaque

FIRST EDITION
*October 2023*

Published from
Joshua Free Imprint – JFI Publications
Mardukite Borsippa HQ, San Luis Valley, Colorado

# SYSTEMOLOGY is
# The New Thought of the 21st Century

It is the study of how Spiritual Beings with
unlimited power became entrapped in the
Human Condition.

This study is an applied philosophy
— "A Pathway to Ascension" —
that charts our way back out of the traps,
freeing the true Spiritual Self to experience
higher levels of existence once again.

In the simplest terms:
Systemology is the true metaphysical science
of the "Matrix."

After more than a decade of development, the
"Fundamentals of Systemology" are concisely
explored here in the very first official
"Basic Course" on the subject ever given by
Joshua Free for the Mardukite Academy.

This collector's edition hardcover includes
all six of the original lesson-booklets for the
"Basic Course" (also available separately).

It's time to discover who you really are...
because you were never "Human."

## The First Official Systemology Basic Course
## on Dynamic Systems of Life and Universes

# MARDUKITE ACADEMY
# "FUNDAMENTALS OF SYSTEMOLOGY"
## BASIC COURSE
### 2023

∞

# EDITOR'S NOTE

"The Self does not actualize Awareness
past a point not understood."
—*Tablets of Destiny*

This book contains a collection of materials from all
six original lesson-booklets developed by Joshua Free
as the first official "Basic Course" on Systemology.

If you read an unfamiliar term not defined in the text,
refer to the "Systemology Glossary" in the appendix.
It is also helpful to keep a quality dictionary nearby.

A clear understanding of this material is critical for
achieving actual realizations and personal benefit
from applying our philosophy as spiritual technology.

The *Seeker* should be especially certain not to simply
"read through" this book without attaining proper
comprehension as "knowledge." Even when the
information continues to be "interesting"—if at any
point you find yourself feeling lost or confused while
reading, trace your steps back. Return to the point of
misunderstanding and go through it again.

Take nothing within this book on faith.
Apply the information directly to your life.

*Decide for yourself.*

∞

*The first official Basic Course on*
*The Fundamentals of Systemology,*
*collecting all six lesson booklets*
*in one collector's edition volume.*

## WELCOME, SEEKER!
## YOUR JOURNEY ON THE PATHWAY BEGINS HERE

This is a basic course in *Systemology*—specifically, the fundamental principles of *Mardukite Systemology.*

Quite simply: *Mardukite Systemology* is a new evolution in Human understanding about the "systems" governing *Spiritual Life, Reality,* the *Universe* and all *Existences.*

In many ways, *Systemology* is a 21st Century break-through that continues the legacy—and unifies the original pursuits—of early 20th Century *"American New Thought"* and other metaphysical schools of philosophy and mysticism. These are mostly all generalized (and often dismissed) in modern culture as *"New Age"* beliefs, though they are actually quite *"old"*—some even based on the most ancient known writings of discovered civilizations.

*Mardukite Systemology* was once concisely described as "an applied spiritual technology of the 21st Century A.D., based on spiritual wisdom from the 21st Century B.C." because of our use of *"Mesopotamian" Arcane Tablets* as source material for its foundations (and from which it retains a *"Mardukite"* designation).

The original *New Thought Movement* in America applied a "Western Civilization" approach to "Eastern" concepts—concepts that we now take for granted today, but of which were relatively unknown to the general population at that time. The movement sought to develop an "applied spiritual philosophy" whereby an individual could unlock their hidden potentials, untapped *"Know-*

*ingness*" and higher spiritual states of *Beingness*. These innate or native conditions of *Self* (as a *Spirit*) are blocked —or "fragmented"—by a "human" preoccupation with identifying *Self* as one and the same with the material body that it is merely using as a "vehicle" to experience (communicate and interact) within *this* Physical Universe.

Early *New Thought* work primarily emphasized practical "healing" applications (*mental healing, faith healing, &tc.*) —but at its very core, we may restate the ultimate pursuit or original focus was to "free humans *to be* their ideal native spiritual state."

This goal has been with us—lingering on the periphery of the "surface world"—for much longer than the existence of a *New Thought Movement*. In fact, for as long as "spiritual beings" have found themselves entrapped by a "Human Condition" and enforced to experience *this* "material existence" (fragmented from their true *Self*), a continuing pursuit has ensued to correct the situation— at least by those individuals still retaining enough *Awareness* to realize it.

Humans have been figuring on how to break free from the "*Matrix*" for a very long time. The desire or ambition to rise above the "standard-issue" Human Condition is already there. But the truth is that many other remotely similar "evolutions" of *New Thought* have dissolved into "multi-level marketing" schemes, "motivational pop-psychology" coaching, abusive "cult-like" movements— or heavily promoted books that skyrocket to the peaks of literary "bestseller lists" only to be discarded soon after and forgotten. They all share one thing in common: they all seem to capitalize on an innate desire or yearning we have to *"ascend"*—but, of course, without delivering stable results.

Even the most pious and well-meaning philosophies and spiritual sciences have each fallen short of piercing the *"invisible barriers"* of perception separating *this* "Physical Universe" from any other "higher" existence—and with it, blocking our "way out" and the *Awareness* of our own true native state as an *Eternal Spirit*.

---

## SYSTEMOLOGY: 21ST CENTURY NEW THOUGHT

Our *Systemology* is a new approach to *"Self-Actualization"*—completely relevant for the modern age and the future—and quite different from previous attempts or other traditions you might find.

Former attempts at overcoming *"barriers"* or *"gates"* of *reality* have included simply pretending that they don't exist, rejecting all material existence—all *time* and *space*—as an *"illusion"* and consequently losing the ability to actually *confront* the *reality* of anything *"As-It-Is."*

Our *Systemology* is also the answer to the "great mysteries" pervading the material sciences and natural philosophies; for they only seek to further qualify and validate the *reality agreements* made for *this* Physical Universe—and thus their level of understanding can never successfully pass the "barriers" either.

When applying our philosophy and techniques, the "systematic routes" outlined for an individual to increase their *"Actualized Awareness"* (and reach gradually higher toward their "*Spiritual Ascension*") is referred to as "*The Pathway*"—and we call that individual a "*Seeker*."

At the start of *The Pathway*, early *routes* emphasize establishing a strong personal foundation of emotional well-being and mental strength before a *Seeker* is introduced to more advanced exercises and practices.

As a *Seeker* increases their *Awareness* in this lifetime, their spiritual "*Knowingness*" also increases—which is to say their sense of "*certainty*"; a certainty on *Life*, on this and other *Universes*, but more accurately, an increased certainty on *Self* as a practically unlimited "spiritual being" *having* an enforced restrictive "human experience."

One of the goals of "*Systematic Processing*" techniques in *Systemology* is to increase the ability of a *Seeker* to actually control and direct the "*attention*" of *Self* as a "spiritual being"—and as a result, *knowingly* increase command of the "human experience." This is a part of what we mean by "*Actualized Awareness*."

---

## THREE STATES OF KNOWINGNESS

Raising a *Seeker's* level of *Actualized Awareness* requires, by definition, "bringing what is *hidden* (or not consciously known) up into the realm of *light* or *Knowingness*." We might go as far to say, as an imperfect example, that there are three primary states of *Knowingness*: *actual knowing, almost knowing* and *not-knowing.*

*Actual knowing* is what an individual is conscious of and can easily recall as needed. It makes up our "surface" (or "above-the-surface") thoughts; what is "*actually known*" and available to *Self* for "inspection" or analytical thought. This includes what we have *certainty* on as part of our *reality.*

Then, there are other *things* "below-the-surface" that we do not easily remember (or have any *reality* on)—and these fit our other categories of *almost knowing* and *not-knowing.* The difference between these other two states is how *far* "below-the-surface" a *thing* is.

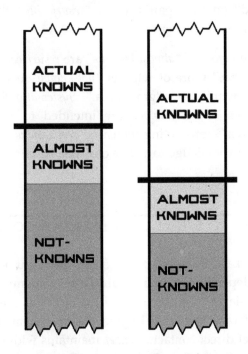

What you "*almost know*" are those *things* just "below-the-surface"—so *close* to the "surface" that they are almost accessible. This "gray area" includes what an individual is *uncertain* of. With a little assistance ("*Systematic Processing*" techniques), you can actually move a *thing* that is "*almost known*" to an "above-the-surface" state of "*actual-*

*ly knowing"* or remembering again. Only then may it be treated with any *certainty.*

There are also memories very deeply buried "below-the-surface." This includes suppressed data that is not currently accessible—and therefore, presently *"not-known."* Once again, there is a way to move *things* from this state into another state. For this to happen, the previous *"almost known" things* ("just-below-the-surface") need to be "purged" (at least partially) by *"resurfacing"* them into *"actually known" things.*

As more layers of *"almost knowns"* are *resurfaced* into *"actual knowns,"* more of what is *"not-known"* becomes accessible within the "gray area." *Systematic Processing* techniques of *Systemology* are intended to target this "gray area"—promoting increased *realizations* by elevating more knowledge to a state of *Actual Awareness.*

---

## HOW TO STUDY A SYSTEMOLOGY COURSE

Most *Seekers* study and practice *Systemology* at-a-distance and independent of the "Mardukite Academy" or any "Master-level" mentors trained therein. This means that the *books* (and to a lesser degree, the *internet*) are the only means of direct contact a *Seeker* maintains with the "Systemology Society" during their studies.

It is quite common to have had negative past experiences with "education" and "learning"—whether in school or other type of instruction. This can sometimes inhibit an individual from pursuing a new *study* later on in their lifetime. However, simply following a few guidelines, ensures a *Seeker's* successful and positive experience when studying this course book—and, of course, the subject of *Systemology* as a whole.

To effectively study and understand a new subject (or a higher gradient of a subject), an individual must be "interested" in the material. A *Seeker* chooses to study *Systemology* because they "want" to, which is to say, on their own "*Self-Determinism.*" While modern society likes to enforce "agreement" (to further solidify a *reality*), a genuine "interest" and true "understanding" can only occur on one's own *Self-Determinism.*

Having established interest, the next *barrier* to understanding is "vocabulary" (words) and "semantics" (meaning). Any specific study, science or tradition is distinguished by the *words* used to communicate it. For true communication to occur, the intended "meaning" for each "word" used must be clearly defined and perfectly understood by the reader or receiver. We call this "*A-for-A*" or "*one-to-one*" communication.

Misunderstood words are the most common reason an individual abandons studying a subject. To relay a proper communication of *Systemology* concepts to a *Seeker*, we use very specific language in our course books. There are newer concepts that more obviously require defining when introduced; and some of our terminology uses familiar words, but with a different or specific meaning than when used elsewhere.

When a misunderstanding occurs, *Awareness* declines. These generally begin to "stack up" after the first occurrence and the level of interest and attention will also decline. This is how a "confusion" develops and the individual will get "bored" with the subject, feel tired, and unable to concentrate.

In extreme cases of confusion, there will be no future interest in studying or "looking at" something  further. Feelings of "anger" and "sadness" may result (because one had originally *intended* on knowing something), followed by lower-level opposing "considerations" such as: "didn't really want to know" or "it probably isn't very good anyways."

The misunderstood word that an individual passed in their study may not be immediately obvious. One solution is to return to the part of the material that was still interesting and enjoyable to read. When scanning around that area of text, there is likely to be a new word (or specific use of a familiar word) that is unclear, but was passed by unnoticed. All *Systemology* books include their own *glossary*. Using this *glossary* and a high-quality dictionary will help resolve this misunderstanding once it is located.

With "interest" and "understanding" secure, the next challenge of learning concerns making a subject "*tangible*"—which means handling it as a "some-*thing*" in the individual's personal *reality* or *Universe*.

Studying intellectual or "philosophical" subjects from a *book* requires excessive amounts of "*thought creation*"—of handling many conceptual images and ideas "*imagined*" solidly in one's "mind" in order to actually "look at" what one is studying. These also require a certain amount of present-time *attention* or *Awareness* to sustain a continual *creation*.

When an individual lacks "objective" examples (objects, graphic representations or direct experience) to examine, they may become "overwhelmed" by "mental-mass" if maintaining too many of their own *images*. This prompts feelings of being "worn out" or "weighed down"—and *considerations* that one "must take a break" or that the subject is "too difficult."

The obvious remedy is to supplement "book-learning" with objective or physical examples. Rather than simply studying or memorizing a series of "dry facts" from an "outside source" (and then returning to "ordinary" life), a student that does understand the material will take it up as their "own" *viewpoint*.

By taking the philosophies up as one's "own" *viewpoint*, the materially is effectively "owned" by the individual. They are not *looking* through a *lens* of someone else. The *"responsibility"* taken by this *ownership* means the freedom to apply information to everyday life and determine the truth of a matter for one's *Self*.

The final *barrier* to learning is the *knowledge* (or "knowledge") itself—the *ledge* or *level* from which a person *knows* or *understands*. A "basic fact" could have many *levels* of potential understanding. To interpret *reality*, an individual "stands" on the *ledge-level* (or *gradient*) of *Knowingness* they have the most "certainty" on.

An effective education of any subject is taught on a *gradient*. This is what is intended by introducing the study of something in *"grades."* Rather than treating a subject as one total mass, true learning is achieved by increasing one's understanding on a *gradual* incline upward. The *ascent* to a mountaintop is not successfully achieved in one leap, but by targeting and reaching specific checkpoints along the way.

In 2019, the *"Grades"* were established for the *"Mardukite Academy"* to properly indicate what level of understanding a specific book or course is intended for. The entry-point to directly study materials of the Systemology Society at the Academy is *"Grade-III."* Lower *grades* pertain to other *Mardukite* subjects treated separately from Systemology. Higher *grades* continue to explore the "theories and practices" of the Systemology Society as a complete *"Pathway to Ascension."*

This *Basic Course* consists of a series of lessons (booklets) that teach the *"Fundamentals of Systemology."* It is an appropriate entry-point for a new *Systemology* student. It is also applicable to more advanced *Seekers* wanting to increase their *certainty* of understanding at higher *grades* as well.

To study *Systemology* just like a student at the Academy: a *Seeker* reads through all instructional material in a *Basic Course* lesson (booklet) and then performs any practical exercises indicated at the end. Before continuing on to the next lesson (booklet), the material is read again and the light exercises are reapplied.

The second pass through the material is likely to result in different *"realizations"* (an increased *level of understanding*) than the first time. Exercises may seem more vivid or significant. *Seekers* should feel cheerful and confident in their *understanding* of a section (or lesson) before proceeding even further on *The Pathway.*

## YOUR FIRST STEPS ON THE PATHWAY

*Systemology* is a "holistic" approach to understanding the human experience. It is not actually a singular "subject" in itself, but rather, a way to "view" the many "subjects" of *Life* and all *Existence*. Its "scope" is not restricted to the rigidly fixed *considerations* of any one "subject" exclusively. Yet, for us to properly communicate its specific intended meaning, *Systemology* does require its own unique basic vocabulary.

The "basic vocabulary" and "*Fundamentals*" of *Systemology* are studied together early on *The Pathway*. They are consistent for the remaining upper-*grades*. It is our *understanding* of them that evolves as we progress.

The entire structure of *Systemology* rests on foundations of earlier material and earlier researches—such as those found in the earlier *grades* of Mardukite Academy. However, in 2019, new developments made it possible for a *Seeker* to start upon *The Pathway* without first spending years navigating around the pitfalls of other avenues and earlier *grade* subjects. As an extension of the original Academy, the Systemology Society continues to map and define the upper-*grade routes* of our philosophy.

The *Fundamentals of Systemology* are explored throughout the *Basic Course*. The critical foundations of its vocabulary and concepts (from *Grade-II*) were concisely collected in 2019 as an essay—"*Mardukite Zuism: A Brief Introduction.*" It is summarized below to provide a more complete introduction to the "lessons" of the *Basic Course*. Each "lesson" will go on to examine this data in greater detail.

## FOUNDATIONS OF SYSTEMOLOGY

*Mardukite Zuism* is a precursor to *Systemology*. It concerns an intensive archaeological study into the *Arcane Tablets* of Ancient Mesopotamia. Such tablet writings were once used to systematize an understanding of all cosmic knowledge—and they include the Babylonian *Epic of Creation*.

The *Epic of Creation* describes *ALL* ("ANKI") as separated into two *existences*: "AN" and "KI"—literally "heaven" and "earth"—which is to say *"spiritual"* ("AN") and *"physical"* ("KI"). Exterior to, and beyond, the *"potential everythingness"* of all *spiritual* existence and *physical* existence is only an Infinity of Nothingness ("ABZU").

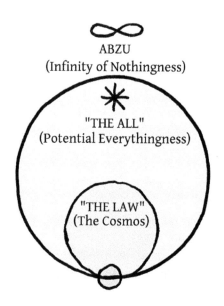

ABZU
(Infinity of Nothingness)

"THE ALL"
(Potential Everythingness)

"THE LAW"
(The Cosmos)

In *Systemology*, we refer to the same two separate states of existence as *"Alpha"* (*spiritual*) and *"Beta"* (*physical*). They are connected only by *"Spiritual Life Awareness"* or *"ZU"*—a term we have retained in *Systemology* (and for which *Mardukite Zuism* is named). Therefore, we have *"spiritual systems"* and *"physical systems"* connected by *"ZU."*

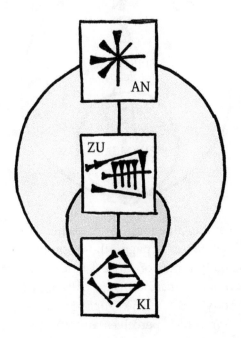

The *"Alpha" Universe*—of *"metaphysical"* or *"spiritual"* energy-matter—is not dependent on the *"Beta" Universe* to exist. The two exist independent of one another, except for a single channel or conduit maintaining a connection, which *is* the *Awareness* (the *Spiritual Life-Energy* or *"ZU"*) of an *"Alpha-Spirit."*

*"ZU"* originates from an *"Alpha"* (*spiritual*) state, separate and distinct from the conditions of *"Beta"* existence that

we experience as the *Physical Universe*. "ZU" is *Awareness* —the *Life-Force* or *Thought-Power* that "acts" or "impinges" on an "organism" in *Beta-Existence*.

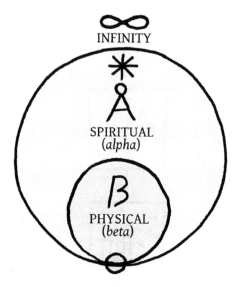

For example: the "intention" to read this book, or "command" a body to turn a page—those specific components are not actually a part of *this* existence. They are manifestations of a *Spiritual Awareness* (*Alpha*) acting upon an "organic body" (*Beta*). The "*Alpha-Spirit*" is the actual "Eternal" *Self*, which perceives and engages with *Beta-Existence* (*e.g.*, "Life on Earth") by using a "temporary" organic body or "*genetic vehicle.*"

The *Alpha-Spirit* engages a "*ZU-Line*"—a *spiritual* "lifeline" of *Attention* and *Awareness* ("*ZU*") energy—to an "organic body" or *genetic-vehicle* in order to directly experience a "*physical*" *Beta-Existence*.

We use the term "*Self-Honesty*" in *Systemology* to describe the original native "*Alpha*" state of true *Self-Directed* "*Be-*

*ingness"* and crystal clear *"Knowingness." Self-Honesty* is the most basic "personality" or true expression of *Self* (*Alpha-Spirit*) as *"I-AM"*—a *Self-Determined* state that is *free* of artificial attachments, automatic reaction-response mechanisms, or enforced (*other-determined*) *"reality-agreements"* concerning the Human Condition.

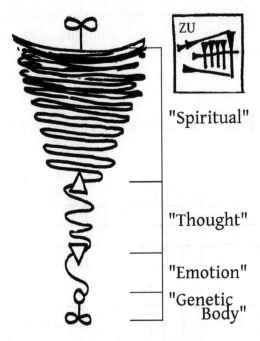

Applying philosophic routes and systematic methods of *Systemology* in order to return *Awareness* of *Self* to its true *"Source"* is referred to as *"The Pathway."* Its structure is based on archaic "models" from the "Ancient Near East" (*Mesopotamia, &tc.*) and elsewhere—such as the *"Chakras,"* the Babylonian *"Ladder of Lights"* (*Star-Gates*), and several versions of *"Kabbalah."*

For example: the Mesopotamians built "stepped-pyramids" as temples—called *"ziggurats"*—serving to remind us of the "ZU" bridging the *spiritual* and *physical* systems. Babylonians constructed *ziggurats* to correspond with *seven* primary "steps" or "Gates."

The "gradients" or "tiers" of the Babylonian *Ladder of Lights* represent *The Pathway*, because they define the *levels* of *Actualized Awareness* (and *Self-Honesty*)—the states of *Self-purification*—between the "standard-issue" *Human Condition* and *Infinity*. And this is the *route* we travel for our "*spiritual defragmentation*" or *Ascension*.

---

## BASIC VOCABULARY REVIEW PUZZLE

## ACROSS

3. The standard-issue default manner of filtering perceptions of the world on Earth, as Self is experiencing it. (2 *words*)

4. The condition of being misaligned, broken apart, shattered, fractured,
distorted, or otherwise separated into parts, compared to its original state.

7. A student or practitioner studying and applying Systemology philosophy.

8. The True Self or I-AM Awareness. (2 *words, hyphenated*)

10. The nature of the Physical Universe or material existence.

11. Another way to say "the agreement about what something is."

## DOWN

1. The physical body, or any organic life, may serve as your ___. (2 *words*)

2. Regimen or routine of Systemology practices, techniques or exercises that increase Actualized Awareness of Self.

5. Returning to the original native state (or Source of the Spiritual Self) is known universally as ____.

6. A stream of energy connecting Spiritual Awareness to physical existence. (2 *words, hyphenated*)

9. The progressive journey taken in Systemology is referred to as "*The ___.*"

# LESSON ONE:
# REDISCOVERING
# THE SPIRITUAL SELF

# LESSON ONE
## REDISCOVERING THE SPIRITUAL SELF

The primary subject of our study in *Systemology* is YOU.

The most important component of all *Life* and *Existence* is the *Awareness*—the *Individual* themselves—that is observing and experiencing it. *Systemology* provides a "systematic" understanding of the parts (or "systems") that make up the *Individual* and the package of experience we call the *"Human Condition."*

A *Seeker* has likely heard of the *"Body, Mind and Spirit"* triad before—whether from "New Age" media or some other "holistic" practice. In spite of *cliché* references in culture, no attempt to effectively understand and command these inter-connected systems has ever been fully completed prior to our *Systemology*.

Humans have spent countless lifetimes questing—*seeking*—for the *"Answer"* to the "great mystery" of *Life* and *Existence*; and ultimately our *Self*. Results of this search have become a massive *confusion;* so much so that many believe that these things are simply not to be known with any certainty. But these things *can* be known about in *this* lifetime.

Previous attempts to understand the connection between *"Body, Mind and Spirit"* have always occurred in *that* order of importance. But "material sciences" have never moved beyond observation of a *"body"*—of the *physical*—and are therefore incapable of perceiving *reality* at any "higher" level of understanding.

31

Even "psychology"—by definition, a study of the "*Mind and Spirit*"—quickly succumbed to evolve as a behavioral neuroscience about the "brain" when its originally defined purpose could not be reduced to a material science. In clinical practice today, it is little more than physiological and pharmaceutical medicine.

The "*Answer*" to *Life* and *Existence* is the *Spiritual Self.* The "mystery" is simply the *confusion*. There is no actual "mystery." However, as long as all attention is placed on trying to understand what's "out there" independent of the individual themselves, the "*Answer*" will not be understood because the right "*Question*" is not being asked.

Our subject is not about some "new" discovery or the domination of some "new" territory—and it doesn't require a "new" age in order to bring it about. It is about reclaiming what we have lost *Awareness* of; what has been *veiled*; what has been *concealed* from view—but it is still right there, because it is the *real* YOU.

*The Answer* is not found by accumulating more levels of *fragmentation* (as other "esoteric traditions" have done). *The Key* is a "sequential" and "systematic" removal of artificial layers—*reality-agreements* that have tracked behind us as we descended from once *knowingly* being a *God-like Awareness*, to now being confined within viewpoints of the *Human Condition.*

"*Rediscovering the Spiritual Self*" is not only the title of this first lesson; it is what defines our progressive journey all along the *Pathway to Ascension*—and, perhaps, is a perfect definition for the entire *Pathway* itself.

Every lesson, technique and exercise in *Systemology* is intended to increase an individual's "*Actualized Awareness*" a little more. The upper-level goal is for *Self* to perceive its own *Beingness* as a actual "*Spirit*"—a viewpoint of an *Awareness* that exists separate from, and exterior to, *this* Universe. This is our true state. To be consistent in our philosophy, we refer to this native "*Alpha*" state of the individual as the "*Alpha-Spirit.*"

The "*Alpha-Spirit*" *is* the individual; it is *you*; it is *me*; it is the "basic personality" of *Self* as a *Spiritual Being* underlying all of the fragmentation of our *Knowingness*. A continuous reduction of *Awareness* has kept us from accessing the virtually unlimited *Spiritual Power* we can *knowingly* command as an *Alpha-Spirit*.

Our *Spiritual Power* (or "ZU") is not actually "lost" to us. It still trails behind us—but it *is* "entangled" in the continuous compulsive creation of our own fragmentation *unknowingly*. Therefore, our *Spiritual Power*, our *Awareness*, and our handling of *Life* and *Existence*, are all interconnected.

By emphasizing our focus on identifying the "*Spirit*" or *Alpha-Spirit*—the actual "*I-AM*" *Awareness*—we immediately discover the original nature of the true *Self*. An individual is not a "*Mind*" or a "*Body*," but a *Spiritual Awareness* with an ability to create and make use of these other "systems"—even if this is happening "automatically" and without *actually* being "aware" of it.

At our source, as an *Alpha-Spirit*, the fact that we continue to exist is not simply a result of our thinking it is so. "*Actualized Awareness*" is more than just: "I think, therefore I know." It is the actual *Awareness* of being *Aware*—being *Aware* of the "thinking" and the "knowing."

We do, however, have the ability to "lower" our "seat" of

*Beingness* (and thereby block our *Knowingness*) so that we experience the "effects" of our *reality-agreements*. This means we have the power to descend from our "*All-Knowing*" state at will, and "cause" our *Self* to "*not-know*" things.

Of course, forgetting that we have done this does create some complications for us. In *Systemology*, we apply our philosophy to correct this—systematically retracing the routes taken by an individual to bring them to their present state. This is done on a "*gradient*" course—gradually and cumulatively handling what is most accessible at each step.

The *Alpha-Spirit* is not actually a "thing" located anywhere in the space-time of *this* Physical Universe (*Beta-Existence*). It does, however, have an ability to locate its own viewpoint by *reality-agreements*. It can change its "seat of consciousness" to relatively "remote" positions —including a "*Mind*" or "*Body*."

There is often a tendency to rigidly fix or "snap-in" one's attention—"point-of-view" or "point-to-view-from"—on the "interior" of, for example, a *Mind-System*; or as in most cases on Earth, inside a "*genetic-vehicle*" (*Body*). This is what we really mean when we say that someone is "stuck in their head" or is "thinking with their (insert body part here)."

Here we begin to see details appear concerning "*Body, Mind and Spirit*" as separate, but interconnected, basic "systems" of the *Human Condition*. And an individual can also "*identify*" so greatly with a "thing" as to confuse its own *Beingness*—going so far as to believe itself to actually be purely *Human* above all else.

# THE HUMAN CONDITION

An *Alpha-Spirit* can command a *"Body"*—or any "system"—directly by "reaching" with "energy-beams." But, as these impulses and intentions became more unknowingly automated, they developed into an energetic-mass (a "mechanistic construct") that we call a *"Mind."*

This *"Mind"* is a communication system (relaying *"ZU"*)—an intermediary used by an *Alpha-Spirit* to control a *"Body."* Of course, in the case of the standard-issue *Human Condition*, the *Alpha-Spirit* also believes that they *are* that *"Body"*—when it is really only a *genetic-vehicle* used to perceive and act in *Beta-Existence.*

The Mind-System may be thought of as an "energy channel" or conduit between the *Awareness* of the *Alpha-Spirit* and the perceptions (and activities) of *Beta-Existence.* The greater the "mass" composing the Mind-System, the less *Actualized Awareness* an individual is experiencing. This "mental-mass" adds "resistance" and "filters" to an otherwise free-flow of energy.

There is a primary part of the "Mind-System" that the *Alpha-Spirit* develops, compulsively creates, and then carries through its own existence. We call this the *"Master Control Center"* (or *"MCC"*) in *Systemology.* Its function is mainly "analytical"—the part we use to "figure about" data and things we experience.

The *Master Control Center* (*"MCC"*) allows the *Alpha-Spirit* to "evaluate" conditions by *differentiating* the data accumulated. Its contents is sorted by *"association"*—that "things" are related to other "things." It is how we "isolate" or distinguish different parts or *"facets"* of an encounter and determine their meaning.

There is also a part of the "Mind System" maintained at a mostly "cellular" level by the *genetic-vehicle*. We call this the *"Reactive Control Center"* (or *"RCC"*) in *Systemology*, because it is primarily a network of stimulus-response mechanisms designed to automatically preserve material "survival" of the organism.

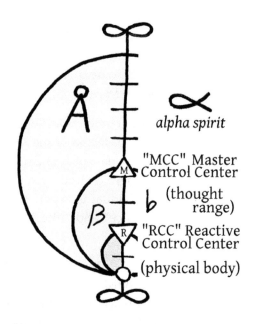

*alpha spirit*

"MCC" Master Control Center

(thought range)

"RCC" Reactive Control Center

(physical body)

The *Reactive Control Center* ("RCC") does not *differentiate* its data. Its contents is sorted by *"identification"*—that "things" are "things." Therefore, if a situation threatens the "survival" of the *genetic-vehicle*, all data from the encounter—any of its aspects or *"facets"*—are "identified" together as equally "dangerous."

Generally, the *Alpha-Spirit* no longer handles "energy" directly in *Beta-Existence*. The *"Mind"* communicates a relay of sensory data and environmental impressions from the *genetic-vehicle* to the *Alpha-Spirit*; and the *Alpha-Spirit* uses it to gauge or estimate the "effort" necessary for intended action in *Beta-Existence*.

Combined, the parts of the *"Mind"* (system) contribute to an *Alpha-Spirit's* experience of the *Human Condition*. If *Awareness* of the *Alpha-Spirit* is "entangled" in *fragmentation* of mental-mass, then there is no clear communication of energy (or data); and therefore, a *"Self-Honest"* experience of *Life* is not possible.

The purpose of *Systemology* as an "applied philosophy" is to research and develop techniques that allow an individual to increase *Actualized Awareness* by *"defragmenting"* the Mind-System. It is not very concerned with the "anatomy" and "physiology" of *genetic-vehicles* directly. The parts of organic systems within *Beta-Existence* are already identified and defined by other "material sciences" and may be studied elsewhere.

---

## BETA AWARENESS

The native state of the *Alpha-Spirit* is "exterior" to *this* Physical Universe and therefore its total capabilities of *Actualized Awareness* extend far beyond the *Human Condition*. When a fragmented "Mind-of-Spirit" (or MCC) and "Mind-of-Body" (or RCC) are combined, the potential range of experience is restricted.

In *Systemology*, the range of perception limited to *Beta-*

*Existence* is referred to as *"Beta-Awareness."* This is what essentially defines the *Human Condition*; because it *is* a "condition" that affects the ideal operations of the *Alpha-Spirit* and other modes of thought and behavior that the individual believes they are originating.

Most individuals assume that they are totally *Self-Directed* and *Self-Determined* in their everyday *Life*—and that they are experiencing *Reality* with clarity. But *fragmentation* hinders this crystal clarity. *Beta-Awareness* is the *Actualized Awareness* (and level of *Self-Honesty*) one maintains in daily activities of a *Human* experience.

The gradients of perceptible *Beta-Awareness* "interior" to *Beta-Existence* and the *Human Condition* (*Alpha-Spirit* + *genetic-vehicle*) are treated in *Systemology* as the *"Beta-Awareness Scale."* This scale directly relates to the *"Standard Model"* of *Systemology*. The *Standard Model* is explored in more detail later in "Lesson 2."

The *Beta-Awareness Scale* is a systematic understanding of varying degrees of thought activity, emotional states, and ultimately, the physical efforts of an individual. *Beta-Awareness* is relatively treated on a scale from "0" to "4." The complete *Standard Model* in *Systemology* extends up to "8"—representing *"Infinity."*

We use the *Beta-Awareness Scale* portion of the *ZU-Line* (*Standard Model*) to "graph" or represent all gradient degrees of *Human* perception between two states: a fully expressed total *Self-Actualized BetaAwareness* (at "4.0") and organic death of the physical body or *genetic-vehicle* (at "0.0").

Both main parts of the Mind-System are also depicted on the *Beta-Awareness Scale*. The *Master Control Center* (MCC) is the point of "contact" between *Alpha* (above "4") and the *Human Condition* (below "4"). Therefore we place the MCC at "4"—and the range of "*Beta-Thought*" *fragmentation* between "2.1" and "4."

The *Reactive Control Center* (RCC) is plotted at "2." It governs the lower gradients of *Beta-Awareness*, which includes many "reaction-response mechanisms" inherent to the development, evolution and survival of organic-physical *Life*. The standard emotional range of the *Human Condition* is between "0.1" and "2."

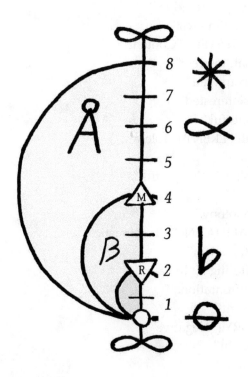

## THE BETA-AWARENESS SCALE

**4.0** SELF-HONESTY (BETA)

3.9 "Vibrant" ("Charismatic")

3.8 "Enthusiastic" ("In Love")

3.7 "Energetic"

3.6 "Cheerful"

**3.5** CONFIDENT ("Positive")

3.4 "Determined"

3.3 "Eager"

3.2 "Alert" ("Attentive")

3.1 "Strong Interest"

**3.0** INTERESTED ("Content")

2.9 "Small Interest"

2.8 "Encouraged"

2.7 "Disinterested"

2.6 "Doubtful"

**2.5** INDIFFERENT ("Tolerant")

2.4 "Bored"

2.3 "Dislike" ("Neglectful")

2.2 "Tired"

2.1 "Monotony"

**2.0** INVALIDATING ("Pessimistic")

1.9 "Antagonism"

1.8 "Suffering" ("In Pain")

1.7 "Confrontational"

1.6 "Violent"

**1.5** ANGRY ("Negative")

1.4 "Hateful"

1.3 "Spiteful"

1.2 "Resentment"

1.1 "Anxiety"

**1.0** FEAR ("Afraid")

0.9 "Terror"

0.8 "Numb"

0.7 "Evasive"

0.6 "Loss"

**0.5** GRIEF ("Sadness")

0.4 "Depression"

0.3 "Victimization"

0.2 "Hopelessness"

0.1 "Apathy" ("Unconsciousness")

**0.0** BETA CONTINUITY (Organic Death)

It is important for us to mention that what is implied by *"emotional"* is really a reference to "negative" states of the *Human Condition*—which are all "reactionary" in nature—such as hopelessness, fear, anger, even pain. We treat enthusiasm, interest, boredom, and such, as "states of mind" or *thought* (above "2").

---

## UNDERSTANDING BETA-AWARENESS

The *Beta-Awareness Scale* portion of the *Standard Model* of *Systemology* shares many similarities with the traditional idea of a "Pyramid of Self-Actualization" (originally proposed by the psychologist, Abraham Maslow). Both models demonstrate a systematic pattern of *Human* "in-the-body" experience and behavior.

As a *Seeker* increases their familiarity with the "*Fundamentals of Systemology*," they discover more about themselves as an *Alpha-Spirit*—and others sharing the *Human Condition*. This type of knowledge leads to "true understanding" that should predict future results or behaviors when properly "evaluated."

Lower levels of *Awareness* relate to states of "emotional turbulence" *encoded* by experiences that are of a "destructive" nature and lead one to reject *Beta-Existence*, often with *force*. Higher levels promote a greater motivation toward achievement and an increased ability to confront and/or change existence "*As-It-Is*."

Obviously, if an individual is experiencing emotional turbulence, they are not "thinking clearly." They are operating from a point-of-view (or "seat of consciousness") that is *beneath* the level of thought and reason.

Between "0" and "4" are various degrees of "*Beta-Fragmentation*" that affect clear perception for the *Alpha-Spirit* when participating with *Beta-Existence*. As an individual becomes more the "effect" of *Beta-Existence*, their level of *Awareness* lowers. An *Alpha-Spirit's* ideal position for involvement in *Life* is at "cause."

*Beta-Awareness* is not only affected by conditions present in one's environment. It is also affected by the *fragmentation* from past-experiences; the "stimulation" (or *resurfacing*) of past impressions (or "*imprints*") that are entangled with specific aspects (or "*facets*") of a situation —and then treated as if they *are* present.

This *fragmentation* affects the clear *Self-Honest* handling of *Reality* by adding false information about present-time conditions. Artificial data is "superimposed" on top of what one is experiencing and agreed to as *Reality*. As a result, the level of *Awareness* lowers, and the individual has more difficulties managing *Life*.

At lower *Awareness* levels, an individual is mainly only seeking and acting in their environment to meet the most basic demands of material survival. Only once these "needs" are met does the idea of greater stability enter in; and with stability established, eventually extending a further "reach" into groups or society—and toward achievement of relative success. This is what the "Pyramid of Self-Actualization" *tiers* demonstrate.

Certainly, *Awareness* levels will fluctuate based on innumerable factors. But, there is also a continuous or chronic state that is maintained in the absence of fluctuation—the "default" position that an individual is likely to operate from most of the time. [This may even differ from the *mask* they try to maintain socially.]

For example: When impressions of past-experience prompt a realization that *loss* is possible, an individual experiences *anxiety* ("1.1"), but *fear* ("1.0") sets in if sensing *loss* is inevitable, and then *terror* ("0.9") once it is about to take place. After experiencing *loss* ("0.6"), if unattended, they succumb to *grief* ("0.5"), and so on.

A *Systemologist* learns to observe the basic patterns of the *Human Condition* so that they do not fall prey to the automatic nature of it. We all experience "ups" and "downs" in the *game* of *Life*—but how "low" we go and how long we remain there indicates just how *Actualized* we have become again as an *Alpha-Spirit*.

Learning to observe the patterns in ourselves and others helps us manage everyday life better. It assists us in understanding "why" we are *feeling* the way we do. It also shines light on "why" another individual is *communicating* the way they are—and even what *behavior* we might expect from them as a result.

Our goal in *Systemology* is to reach a level of *Self-Honesty* that allows us to rise above restrictive viewpoints of *Beta-Existence*, to realize just how much "*more than Human*" we really are, and to know with utmost certainty that there are much "higher" *Universes*—only forgotten —for us to *return* to *again* as *Alpha-Spirits*.

## PRACTICE EXERCISES

1. Can you recall a pleasant moment? What happened? How long ago was it? Where did it take place? Who was there? What did you like about it? You may have heard the expression: "think a happy thought." At times when a person is feeling emotional turbulence or stressed, the quickest remedy is to *locate* a pleasant incident and *notice* something about it. When an individual becomes overwhelmed or confused, it is generally helpful to go back and work from a point of certainty or "stable knowingness." Should you find any such difficulties with *Systemology* studies or exercises (or elsewhere in everyday life), simply apply the suggestions from this first practice.

2. Look around the room. Locate an object. Notice something about it. Do this with several objects. Notice many things about the same object. Identify only what you can actually *see* about these objects. Make no assumptions, computations or considerations otherwise. Continue this practice until the room seems "brighter." Many individuals associate the concept of *"Awareness"* (or "being aware") with the idea of someone "knowing" (with certainty) what is going on around them and being "observant." This is quite accurate. The ability to "observe clearly" is a practiced skill. Only by confronting existence *"As-It-Is"* will we have our full power of *Awareness* to affect it.

45

3.  Consider some individuals who are (or have been) influential in your life. With each one, *spot* them on the *Beta-Awareness Scale*—which is to say, *locate* the position on the scale that they seem (or seemed) to be operating from consistently. What did you *notice* (or *recall*) about the individual that resulted in your decision? How are you similar to this person? How are you different? Often times, we will take on characteristics and personality traits of those closest to us—especially those we admire or consider an "ally" to our continued survival. It even happens out of "sympathy" for those we have lost. An individual may adjust their own "tone" or "energy" to be in better communication with a person operating at a certain *Awareness* level—so long as they are *knowingly* doing so. *Fragmentation* allows the "basic personality" of the *Alpha-Spirit* to be affected *unknowingly.*

4.  Go to a public place where you can directly observe other people performing activities and/or interacting. Practice your "observation" skills and understanding of the *Beta-Awareness Scale.* When you *observe* an individual, *notice* things about them—facial expressions, body position, &tc.—and *spot* them on the scale with your best estimation. [Keep this evaluation to yourself.] If your situation allows for you to continue observing afterward, see if any further communications or behaviors prompt you to change your opinion. Practice observing different individuals this way.

5.  Locate a "neutral" object—one that you do not have strong feelings of *liking* or *dislike* toward. Get the sense of feeling *indifferent* about it—actually deciding or *intending* to be *indifferent* toward it. Decide to now be suddenly very *interested* in it —and get the sense of actually being *interested* in it. Now, decide to be *afraid* of it—and get the sense of really feeling that. Choose various positions on the *Beta-Awareness Scale* for practicing this. End off on the same "neutral" state you started with. An *Alpha-Spirit* may certainly decide to *knowingly* experience an entire range of potential sensations, feelings or emotions. But, *fragmentation* causes these reactive-responses without *Self-Determination*.

6.  Locate a "neutral" object—as with the previous exercise. This time, get the sense of *intending* for the *object* itself to genuinely "feel" the various positions on the *Beta-Awareness Scale*, from its viewpoint. Each time, get the idea that the object is "feeling" this reaction toward you—so, *afraid* of you, or *interested* in you, or *angry* at you, &c. Does confronting any of these states change the way you feel? Do you make your intentions as a "silent command"? Or, does a certain "*image*" come to mind for each state? Maintaining one's own "inner composure" is the key to remaining at "cause" while confronting the nature of *Reality* "*As-It-Is*." The less we are reacting "on automatic," the more *Actualized Awareness* we can apply to managing our own *Self-Determined* existence.

7. The final exercise of the first lesson is to apply an understanding of the *Beta-Awareness Scale* and *observe* your *Self.* Have you spotted your own chronic state on the scale? Keep in mind—we do not mean how things are when watching your favorite movie, or getting a visit from an unwelcome guest. We are interested in the underlying continuous position from which we view the world. In addition to this, we *do* want to observe the fluctuations—and take notice of the circumstances that surround this. We also want to make records of these changes; how far they shift and for how long.

# LESSON TWO:
# SPIRITUAL LIFE AND
# THE UNIVERSE

## LESSON TWO
## SPIRITUAL LIFE AND THE UNIVERSE

Our previous lesson (booklet) introduced the nature of the *Alpha-Spirit* and the basic parts involved with experiencing the standard-issue *Human Condition*. In this lesson, our focus shifts from the subjective "view" of *Self*, onto the objective environment that *Self* is "viewing" — and the "*reality*" that is being *agreed* upon.

"Spiritual techniques" of *Systemology* apply the philosophy of our *Standard Model* — which is based on the "*Arcane Tablets*" that were lost to us for thousands of years. These archaic teachings form the background of the modernized graphic depiction (and philosophy) of the *Standard Model* that we use to research practices.

The most ancient writings of cosmic wisdom essentially state: "The ALL is composed of the '*Alpha*' Spiritual Existences and the '*Beta*' Physical Existences, which are divided (or separated) by '*Cosmic Law*' and connected by a '*Spiritual Life-Awareness*' ('*ZU-Line*') — beyond which can only be an Infinity of Nothingness."

Many *Seekers* have some innate sense that while there *is* a Universe — a "world" — around us that we can *see*, this is not the "whole" of existence that we are experiencing. And by this, we do not mean other "locations" *interior* to this *Beta-Existence*. However, we can only experience what we can perceive — and those relatively higher "Gates of Perception" seem veiled to us while we are still struggling to handle the *Human Condition*.

As stated in the previous lesson: the *Alpha-Spirit* no longer practices directly handling "energy" while tied to *considerations* and *reality-agreements* of *Beta-Existence*. This includes handling energies *inside* this Universe. Sound, pressure waves, light and other particles are all "sensed" by the *genetic-vehicle*, not the *Alpha-Spirit*.

Data from external sources is communicated back to the *Alpha-Spirit* through "relays" of the Mind-System. A "*ZU-Line*" connecting an *Alpha-Spirit* to, for example, an organism in *Beta-Existence*, is also the conduit or channel for communicating information between. *Fragmented* communication leads only to *confusion*.

If *Actualized*, the *Alpha-Spirit* does have an ability to *knowingly* use a "body" (a "remote point-of-view" from itself) in order to interact with, and experience, any Universe, simply by changing *agreements* about *Reality*. However, we have to own or take responsibility for such a "choice" for us to have any power to change it.

In the case of the standard-issue *Human Condition*, there is no "surface memory" of our making *reality-agreements* —and therefore, no *knowingness* to properly change them with "high power" *Awareness*. An individual must first *spot* and *realize* the point which they *actually* made a decision in order to have any *actual* intention to *change* it. How else can we change a decision we don't understand ever having made?

The *Standard Model* is the chart of our "descent" of *Awareness* as *Spiritual Beings*. It is reflected in the lore of most mystical, spiritual, and religious, teachings—yet it is never perfectly communicated or understood. By applying *Systemology*, this chart may be used to effectively reverse directions of our spiritual evolution.

# INTRODUCING THE STANDARD MODEL

The *Standard Model* of *Systemology* provides us with a foundation to explore some of the most ancient *"axioms"* of *esoteric* philosophy on a "practical" level. Seek and you will find. To know thy own *Self* is to know the Universe. As above; so below. As within; so without. As the Universe; so the *Spirit*...

Although simple in its graphic detail, the *Standard Model* required a decade of development before the first portion of its interpretation appeared as *"The Tablets of Destiny Revelation"* and *"Crystal Clear"* in 2019. All of the techniques and applications of our philosophy are actually derived from this understanding.

Originally the *Standard Model* only consisted of background models—spheres, circles, levels and *"Gates"*—reflected in the Mardukite (Babylonian) "Ladder of Lights," the Druid's "Cabala, and other *"cosmological"* *paradigms*. The *"ZU-Line"* concept developed independently. Only once we combined them did we discover an effective workable *methodology* for what was first theorized as *"Systemology: The Original Thesis"* in 2011.

On paper, it is quite basic: the *"ZU-Line"* (a spiritual or *"Alpha"* continuum of an individual's *Awareness*) intersecting various concentric circles (representing complete "systems" imposing certain conditions at different levels or gradients). Where *Awareness* "meets" environment, we have the experience of a *Reality*.

The *"ZU-Line"* is a stream of *Awareness* of the *Alpha-Spirit* —the individual as *Self*, the spiritual "I-AM." The *Alpha-*

*Spirit* is not located in space-time, but it may operate a "point-of-view" potentially anywhere. Therefore, the *"ZU-line"* is a *"continuum"* that stretches across any and all background models of *Universes*.

Each of the circles or spheres represent a complete *"system"*—what we might even go as far to call a *"Universe."* At their own *level*, they are each a "completion" or "totality." That is not to say that they do not also in some way affect, or are affected by, other "sealed" *systems* (that only seem exclusively independent).

The study of the *Alpha-Spirit*, the *Mind*, the *genetic-vehicle*, and *Beta-Existence* can be best conducted by treating each one as a *Universe*—as a *system*. On the *Standard Model*, each circle—each *Universe*—also has its own *"continuity"* or "zero-point." This is also a theoretical point of communication to a "lower" system.

For *Beta-Existence*, we mean the point in which energy "condenses" into its most solid form of matter (or *inert matter*) for that level of existence. It may also be found at the point in which energy-matter collapses, such as with a *"black hole."* If an action takes place, it is the point where things are relatively "at rest" again.

For the *genetic-vehicle* (or *Alpha-Spirit* + *genetic-vehicle*), it is experienced at the point below *"apathy"* on the *"Beta-Awareness Scale"* [see *"Lesson 1"*] as *"organic death."* This state is reached when the *genetic-vehicle* is so overwhelmed (is the *effect*) by *Beta-Existence* that the *Alpha-Spirit* ceases to maintain its *"ZU-Line"* directly. Cellular function ceases in the biological organism, but an *Alpha-Spirit* continues its own spiritual existence.

Above "0" on our scale (or model), we have the *Spiritual*

*Awareness* of *Life* in *Beta-Existence.* The lower levels pertain directly to the biochemical functions of the organism (*genetic-vehicle*) itself.

Between "0.1" and "2" we treat what is generally referred to as *"emotion"*—and *emotionally encoded* reaction-responses to data.

We consider the *emotional* range as its own *system,* because it functions like an independent Mind-System, but one that is specifically connected to *Organic Life* in *Beta-Existence.* It does not "analyze" or "reason"—but functions solely on *stimulus-response* mechanisms.

In "archaic" *Systemology,* this faculty of the *genetic-vehicle* was called the "Mind-of-Body." In today's *Standard Model,* we position the *"Reactive Control Center"* (or "RCC") at "2" on the *ZU-Line.* It governs the relay of communication between the *genetic-vehicle* and higher levels of the larger Mind-System.

Between "2.1" and "4" we treat analytical levels of attention and what is generally referred to as *"thought."* This includes portions of the Mind-System that the *Alpha-Spirit* has direct command over, so it was once referred to as the "Mind-of-Spirit"—because it is a higher faculty that is combined with the lower system.

This "archaic" interpretation is not our best understanding, because an *Alpha-Spirit* is accustomed to using the whole of the Mind-System to experience the *Human Condition.* However, we position a *"Master Control Center"* (or "MCC") at "4" on our model, indicating a point of contact between the *Alpha-Spirit* and *"Mind."*

In this wise we are able to distinguish that an *Alpha-Spir-*

*it* is operating remotely (from an *Alpha-Existence*) and *exterior* to *this* Universe (*Beta-Existence*). Through fixed attention of *Awareness*, *Self* is able to experience the *Human Condition*—and its interface or control point of a *genetic-vehicle* is the MCC of the Mind-System.

The MCC—or analytical system of "*thought*"—influences, and is influenced by, the lower-level biochemical system of "*emotion*." In turn, this system of "*emotional*" communication is what directly animates a "*body*." It is governed by the RCC, which also filters all sensory information communicated back to the *Alpha-Spirit*.

On the *Standard Model*, the boundaries of the *Human Condition*, the "Mind-System," and *Beta-Existence*, is position "4." This range of "0"-to-"4" on the *Standard-Model* is "synchronous" with the *Beta-Awareness Scale* [in *Lesson-1*]. Relatively, everything above "4" is treated as *Alpha*; that which is not tied to "being human."

In *Systemology*, a *Seeker* regains more direct command of the Mind-System (or MCC), and thereby a greater control over the related sub-systems of the Human experience. *Self-Honesty*, as it relates to *Beta-Existence*, means having a clear "*defragmented*" channel of communication between the "*Mind*" and *genetic-vehicle*.

The position of "8" is always representative of *Infinity*—or that which is beyond what we can identify as the native original state of the *Spiritual Self* (or *Alpha-Spirit*) at "7" on our *Standard Model*. This is the level of one's own "personal" *Home Universe*. And here we see the ancient "seven-stepped pathway" to *Self*.

Technically, an individual has never actually "left" its native state; but it can *consider* that it has. Not actually being a locatable position an *Alpha-Spirit* may potentially *consider* being as *any* position. Of course, those *considerations* become more limited as one takes on more *layers* of "personality" and "character."

When fixated on the *Human Condition* and *Beta-Existence,* the *Standard Model* is able to chart the various "positions" of *Beingness* that an individual is considering *Self* to *be*. The lower the level of *Beta-Awareness*, the lower the point of *Beingness* that is being determined *for* the *Alpha-Spirit,* whose own true *Beingness* and ideal state of *Awareness* is not positioned *interior* to *Beta-Existence*, but from an *exterior* point-of-view.

The remaining *Alpha* domain on our *Standard Model* is reserved for the subjects of *"Alpha-Thought"* —and the "creations" and "games" that an individual engaged in preceding the *Human Condition* and *Beta-Existence*. It still relates to a Human experience in terms of *"intention"* and *"imagination"* —which are *Alpha* qualities.

# THE STANDARD MODEL OF SYSTEMOLOGY

### – – *ALPHA* – –

8. "INFINITY"

7. The Alpha-Spirit (*Home Universe*)

6. Alpha-Thought (*Creation and Imagination*)

5. Will-Intention (*Alpha Equivalent of Effort*)

### – – *BETA* – –

4. "MCC" (*Command of a Mind*)

3. Analytical (*Associative Knowledge*)

2. "RCC" (*Control of Biochemical Activity*)

1. Emotional (*Stored Data of Loss and Pain*)

0. Effort (*Solidity of the Physical Universe*)

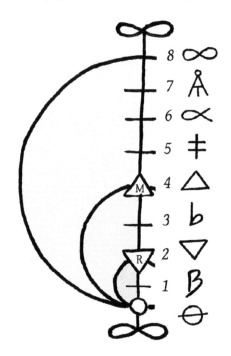

# THE SPHERES OF BETA-EXISTENCE

The *Standard Model* takes an "objective" view of *all* potential positions of *all* potential *Universes*. It is an illustration of "spiritual descent," but also a map to chart the way back toward *Ascension*. Additional work with this model resulted in the development of an alternative "subject" view called the *Spheres of Existence*.

It has been suggested in our philosophy that the *Alpha-Spirit* departed from their own original native state in order to "create" with other *Alpha-Spirits*—or else to shift *Awareness* from a personal *"Home Universe"* to a *"Shared Universe."* The *Alpha-Spirit* is still very much operating from a "personal universe"—however, in this case, the individual *alters* it to also simultaneously "share" the *reality* of other entities by *"agreement."*

As an individual takes on more layers of *agreements*— and *fragmentation*—the qualities and considerations of a "personal universe" continue to "condense"—or "collapse"—into lower levels of *Knowingness* and *Beingness*. The apparent result of this "descending spiral" has led us to such a state as the *Human Condition*.

From the perspective of experiencing the *Human Condition* in *Beta-Existence*, the solution to our *Ascension* is to "free up" the entanglement of our *Awareness* and *considerations*; to "extend" rather than "retract" our *"spiritual reach"*; to rehabilitate the native *Self-Honest willingness* to confront all potential existence *"As-It-Is."*

The *Systemology "Spheres of Existence"* chart allows us to

systematically gauge this *"spiritual reach"* from the perspective of the *Alpha-Spirit*, operating as a *genetic-vehicle* in *Beta-Existence*. *Alpha-Spirits* operate toward some "goal"—as in a *"game."* And the common denominator of *"games"* in this *Beta-Existence* is to *survive*.

Ideal activities and "goals" of the *Alpha-Spirit* are all *creative*. Hence, the most optimum level of *survival* in *Beta-Existence* is achieved by effectively applying *Alpha* qualities—*imagination, reason, ethics* and *aesthetics*—to the Human experience. Even prior to reaching *Ascension*, we can improve qualities of *Life* here on Earth.

Our *"Spheres of Existence"* philosophy only partly aligns with the *Standard Model*. It begins with *"Self"* at "1" and extends outward into "existence" from there—as does our *reach*. At "1" we identify the individual as an entity in *Beta-Existence* operating a material "Body." The individual is a singular unit of *Spiritual Awareness* (or *ZU-Line*) "impinging" on the Physical Universe.

This philosophy can apply to *any* individual "lifeform" or "consciousness" on *any* planet of *Beta-Existence*. It is simply the case that we are accustomed to viewing such metaphysical models—the Babylonian *Ladder of Lights*, the Druid's *Cabala*, the *Chakras*, *&tc.*—from the perspective of the *Human Condition* on *this* planet.

In "archaic" *Systemology*, we referred to this model as "Circles of Influence." Each circle is a "system" unto itself, and they do mostly stretch "horizontally" across *Beta-Existence*. However, they are better depicted as encompassing (or surrounding the whole) of one another as a series of "spheres" within "spheres." As they represent the nature of *this* existence and the game-goal *"to exist,"* the newer terminology is more accurate.

# THE SPHERES OF EXISTENCE MODEL

**– – *METAHUMAN* – –**

8. "INFINITY"
7. Alpha (*All Life in Spiritual Existence*)
6. Universe (*All Life in Beta-Existence*)
5. Planet (*All Life on Earth, Trees, Animals...*)

**– – *HUMAN* – –**

4. Species (*All Humans*)
3. Societies (*Groups, Organizations*)
2. Home (*Domestic, Family, Children*)
1. Self (*The Individual*)

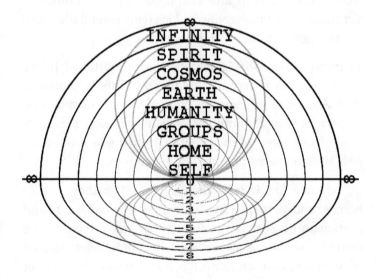

## REACHING THROUGH THE SPHERES

There are many goal-oriented purposes for the activities of *Spiritual Life*. However, in regards to the *Human Condition (Alpha-Spirit + genetic-vehicle)*, the Primary Directive or motivation pertains to material survival. When we say *"to survive"* (or *"to exist"*) it does not simply mean barely scraping by just to see another day. However, an individual that *is* "surviving to see another day" must at least be evaluating *something* right.

The *Spheres of Existence* are *"existential."* They pertain to the experience of conditions *in* existence—domains that support, and are supported by, various systems of *Life*. Many individuals fix attention on one specific domain—yet these are all interconnected systems essentially equal to one another to support *Actualization*.

In many ways, *Self-Actualization* is proportional to the ability to "reach across the spheres." This is similar to how *Actualized Awareness* is equivalent to how much *defragmented* clear communication is relayed across the *ZU-Line* of the *Standard Model*. Data from these models assists in effectively applying our techniques.

Individuals are indeed themselves ("1"); they live in homes, and perhaps have a family ("2"); work jobs and participate in societal groups ("3"); and, of course, are members of the *human* species ("4"); one of many species of this ecosystem on Earth ("5"); and so on. What is not generally equal is how *Self-Honest* or *Actualized* an individual actually is in their clear understanding and communication at all of these *levels*.

Each of these *Spheres of Existence* is composed of many specific aspects or *"facets"*—some of which are likely to have more "turbulent" energy entangled up in them than others. Any such associated turbulence—or *fragmentation*—affects the clear handling of these *facets* of existence and achievement of total *Actualization*.

An individual generally develops a "personality" of inclinations or patterns based on stored experiences. The tendency to "reach toward" or "retract away" from one or another *facet* becomes automatic. While an *Alpha-Spirit* is certainly entitled to prefer one or another interest, automated tendencies are not very *godlike*.

Those that are not *Actualized* at the first sphere of *"Self"* may still continue to operate a *fragmented* reach out to other spheres by their interconnected interaction. One person may believe raising a family is the most important thing, while another neglects theirs for a career, or to save a forest. They are seldom all balanced.

The *Spheres of Existence* model also plays a significant role in upper-grade studies regarding *"Utilitarian Systemology"*—which is to say, *ethics*. In brief: the "greatest good" or "ideal course of action" is that which helps, promotes, or assists, continued "optimum survival" at the highest sphere or *Life-System* of existence.

When we reactively *"withdraw"* from some *facet* of *reality*—something we've already *agreed* is *real*—then we are not willing (and therefore "unable") to *confront* it *"As-It-Is."* These "reactive mechanisms" are a source of *fragmentation*, because they still require part of our *Awareness* (or energy) to continue beneath the surface.

The key *realization* for a *Seeker* to maintain while pursuing their *Systemology* studies is that at our core—our native state—we are *Spiritual Beings* with essentially unlimited power to *create*. The track of our spiritual existence is an accumulated series of descending considerations and *reality-agreements*.

Systematically, we must work our way out to uncover the secret of how an *Alpha-Spirit* allowed itself to be entrapped in these ongoing *games* of *survival*, when the one thing an *Eternal Being* cannot help but continue doing, *is* survive.

## PRACTICE EXERCISES

1.  While seated in a chair, *get the sense* of You "mak-
    ing" *that body* "sit" in a chair. [Practice this sever-
    al times or until you have a *reality* on it being so;
    then continue reading.] Perform the same basic
    exercise, this time applying it to whatever the
    "body" is doing. For example: if you are holding
    this book, then *get the sense* of You "making" *that
    body* "hold" this book. The *Human Condition* is ac-
    customed to frequently functioning on "automat-
    ic" reactive-responses. This applies not only to
    cycles of thought and emotion, but also the pat-
    tern of our behaviors. We are often not in a habit
    of thinking, feeling, and acting, *deliberately.*
    Therefore, take some time each day to *knowingly*
    command the Human experience by practicing
    this exercise. At first, you may only maintain a
    few minutes of *conscious control* before attention
    drifts. Eventually, you will be able to increase the
    duration you can hold this concept, and extend
    your application of this "sense" to more and
    more of everyday *Life.*

2.  On a blank sheet of paper, draw out your own
    version of the *Spheres of Existence* model. Begin
    with the smallest circle—labeling it "SELF"—and
    then proceed to draw the other seven circles
    around it, using the list from this lesson to label
    them, too. Inside the space you've provided to
    represent each "*Sphere,*" list as many words—as-

pects or *facets*—that you "associate" with that specific domain. It is most beneficial to *identify* actual "things" or "beings"—with "mass"—but feel free to include any "concepts" that also surface. If space on your model becomes too restrictive and you are having to be too general, you can continue your listing on separate sheets of paper. Some examples from other *Systemology* materials include: ("1") *Self, Your Body*; ("2") *Home, Family, A Specific Relation, Sex, Children*; ("3") *Groups, Community, A Certain Group, A Certain Type of Person*; ("4") *Human Species, Nationalities*; ("5") *Planet Earth, Animals, Nature*; ("6") *A Solar System, Galaxies, Physical Universe*; ("7") *Spirits, Entities, Mysticism*; ("8") *Infinity, God, Religion*.

3. Start a notebook to keep records of your practices as written exercises. Record any data from the exercises and any realizations that occur as a result. Using the list you prepared in the previous exercise: consider (or reflect on) what your feelings are concerning each listed item. What items do you prefer and which do you dislike? Recall individual times when you had encounters with each item. What have you done about (or toward) the item? What has that item done to you? What have you observed others do about (or act toward) the item? How has the accumulation of experiences (regarding a particular item) affected your inclinations to either like or avoid it. How does having a particular "belief" or "feeling" about an item affect what you are now willing to experience?

4. Spend some time outdoors—or looking out through a window—and see how many forms of *Life* you can *identify* as existing in your surrounding environment. Take notice of the things you can actually *see*. In each case, consider (or reflect on) *"what"* *Life* is actually *doing*. Then consider what *Sphere of Existence* it best applies to. Finally, consider what relationship it has with other *Life*: How is its continued existence supporting, or being supported by, other *Spheres of Existence? Systemology* is named for its approach of studying existence as a composite of living or "dynamic" (variable) *systems.* Humans are accustomed to treating *reality* as both solidly fixed *and* fragmented into a myriad of unrelated parts. By looking closely and fixedly from one "point-of-view" we exclude all others. We forget that everything we *see* is interconnected—even connected to levels of existence that we are not presently perceiving. *Self-Actualization* increases with the range of existence that an individual is *willing* to (but is not required to) extend its *"spiritual reach"* toward and experience.

5. Go to a public place where you can directly observe the activities of *Life* and directly notice various things around you. *Identify* specific examples of each of the *Spheres of Existence* in your surroundings. You may begin with the *Lifeforms* and various "things" that you first *notice*, and then assigning their place on the model. Secondly, you may take each category (or existential system) in turn and *spot* the items in your environment that apply to that specific *Sphere of Existence.*

6. Taking individual *Spheres of Existence* in turn, *imagine* the world from the "point-of-view" of each one exclusively. Record any *realizations* that you have while performing this exercise.

7. Locate an object in your environment. *Decide* that you will *reach* for the object; then *decide* to "make" the "body" do so—and finally *do* it. Now, *decide* that you will *let-go* of the object; *intend* to "make" the "body" do so—and then *let-go*. Practice this exercise of *knowingly* locating, then "reaching to touch" and "retracting your touch," for different objects. Then, choose a large enough object that you can practice this exercise repeatedly for different *spots* on the same object. The *Human Condition* is full of "automatic mechanisms" and "reactive-responses" that seem to affect control of the "body" and place the *Alpha-Spirit* at an "effect" point, rather than "cause." Tendencies to *hold onto* or suddenly *withdraw from* various aspects, persons, or *facets* of *Life*, often take place outside of one's conscious (or *Actualized*) *Awareness*. Although we might apply "reason" or "justification" after the fact, the truth is that the *Alpha-Spirit* operates best from a "defragmented" state of *Self-Honesty*—fully *Self-Determined* in all of its creative expressions, directed attention, and intentions for action.

# LESSON THREE:
# THE FILTERS OF
# HUMAN PERCEPTION

## LESSON THREE
## THE FILTERS OF HUMAN PERCEPTION

*"If the doors of perception were cleansed,*
  *every thing would appear to man as it is,*
    *Infinite.*
*For man has closed himself up,*
  *until he sees all things*
    *through narrow chinks of his cavern."*
                    —William Blake (1793)

Many individuals are likely to define *"perception"* and *"Awareness"* as the same—and they *are* quite similar. The basic range of "Human Perception" is charted on the *"Beta-Awareness Scale"* [in *"Lesson 1"*] and between "0"- and-"4" on the *"Standard Model"* [illustrated in more detail in *"Lesson 2"*]. But there is more to know about in *Systemology* than is displayed in the graphic models of these previous lessons (booklets).

As we experience it, *"perception" does* involve *Awareness*— because it concerns information ("data") that the *Alpha-Spirit* (the individual or actual *Self*) is *"aware of."* In regards to the Human Experience: what we are *"perceiving"* relates to "sensory data" about the environment (usually restricted to *Beta-Existence*)—or about the condition of the *genetic-vehicle* itself. By "sensory data" we mean *sensed* using "physical body" *sensors.*

We differentiate the "Human" range of *perception* in *Systemology,* because it relates to communication-relay between the *"Body"* and *"Mind."* It is only as a result of this *internal "processing"* of sensory perception by the

Mind, that the *Alpha-Spirit* is *"aware of"* any Human Experience—what a *genetic-vehicle* is experiencing.

For example: if we speak of the *senses* or *sensation*, we mean the direct "stimulation" of *sensory receptors* of the physical organism in *Beta-Existence*; whereas *perception* is the assigning of meaning to sensory data. The *"Body"* transmits signals to the *"Mind"* for interpretation. This is the extent that *Self* can be *"aware"* without *actually* "looking." A *fragmented* unclear communication on this line results in a difficult experience for *Self.*

The philosopher, Aldous Huxley, mentions his own belief about *Self-Honesty* in his book *"Doors to Perception"* (1954), titled after the William Blake quote opening this lesson. He refers to it as the "Mind-at-Large," and cites his academic inspiration, quoting a Cambridge philosopher, C. D. Broad:

> "Each person is at each moment potentially capable of remembering all that has ever happened to him and of perceiving everything that is happening everywhere in the universe. The function of the 'brain' and 'nervous system' is to protect us from being overwhelmed and confused by the mass of largely useless and irrelevant knowledge, by shutting out most of what we should otherwise perceive or remember at any moment, and leaving us only that very small and special selection which is likely to be practically useful. According to such a theory, each one of us is potentially Mind-at-Large."

Later on, Huxley describes his own pinnacle *realization* of what he considers *"egolessness"*:

> "In the final stage of [*Self-Actualization*], there is an

'obscure knowledge' that All is in all—that All is actually in each. This is as near, I take it, as a finite mind can ever come to 'perceiving everything that is happening everywhere in the Universe.'"

As an applied philosophy, *Systemology* techniques are intended to systematically "clear" or "cleanse" the "lenses" of *perception*. For the *Alpha-Spirit* to achieve any degree of *Self-Honest* "point-of-view" (or "POV") on the Human experience, we systematically examine and "analytically process" various layers and degrees of *fragmented* thought associations, reactive-response mechanisms and emotional turbulence.

---

## THE NATURE OF FRAGMENTATION

We have introduced the nature of the *Alpha-Spirit*, or individual themselves [*in "Lesson 1"*]; and the nature of the *existences* an individual experiences as *reality* [*in "Lesson 2"*]. The next main focus of study is called *fragmentation*. It regards an *Alpha-Spirit's* clarity of *Awareness* and certainty when "perceiving" any *existence*.

True *Knowingness* is an *"Alpha"* quality. It seldom enters into the *Human Condition*, which is "wired" to excessively "figure" and "think." In the absence of *Knowingness* and certainty (at *"Alpha"* levels), we likely experience a network of associative *thought* or *"Thinkingness"* (and *fragmentation*) within the Mind-System.

At an *Alpha* level, when a *Spiritual Being* is "interested" in something, it *knows* by *looking*. If experiencing the *Human Condition* (*Alpha-Spirit* + *genetic-vehicle*), and that "point-of-view" (POV) is extended toward *Beta-Existence, Awareness* passes through various systems of *thought* and *feeling* (as represented by the *ZU-Line*).

Our thoughts and emotional turbulence have the ability to affect our *perceptions*—our experience of *reality.* The *ZU-Line* represents a metaphoric "telescopic *lens*" that the *Alpha-Spirit* uses to *perceive* "interior" *Beta* conditions from an "exterior" *Alpha* source-point. It has many layers of potential filtering or *fragmentation.*

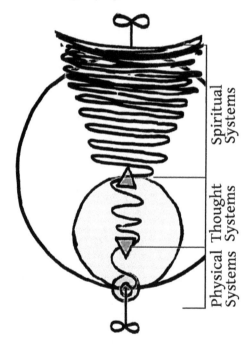

Spiritual Systems

Thought Systems

Physical Systems

One of the primary issues with experiencing *Beta-Existence* using indirect relays of communication, is the *fragmentation* inherent in the Mind-System. It acts as a relay of *perceived* data between the *genetic-vehicle* and the *Alpha-Spirit.* The relay from the *"Mind"* to the *Alpha-Spirit* is at "4" (the "MCC") on the *Standard Model.*

The same message is conveyed to the *Alpha-Spirit* whether *sensed* from an *external* environment in *Beta* or from the *internal* mechanisms of the *"Mind."* The Mind-System does not distinguish between "internal" or "external" sources of *perceptual* information. Determining *"source"* with certainty is a part of *Self-Honesty.*

For example: whether something *sensed* in the external environment is an "actual" presence of danger (or emergency), or whether environmental *facets* are "only" internally *associated* with an idea of danger (due to *fragmented* data from past experience), the same *reality* is *perceived*: that "environment equals dangerous."

The first type of personal *fragmentation* studied in *Systemology* is called *"imprinting."* An *"imprint"* might be best understood as a *"slate"* or *"glass slide"* storing emotional energy from turbulent memories. These *slides* replace, obscure, or distort, the data *perceived* by an *Alpha-Spirit*—and its "clear view" of experience.

To understand better what *imprints* are—or what an *"imprinting incident"* is—consider the typical definition of the word *"imprint"* itself, prompting our choice of using it. It is a "strong impression" or "facsimile image" of something *perceived* as significant. Consider, for example, your "first impressions" of something.

For example: there are many "words" that when spoken seem to have a "trigger" effect on an individual. There is something reactively summoned into view of the *"Mind"* that is then responded to automatically. We refer to this type of information as *"imprinted"* upon the individual. Regardless of what is presented or experienced afterward, the entire "subject" or "concept" will be "filtered" through unclear *imprinted* lenses.

## EMOTIONALLY ENCODED IMPRINTS

In addition to the *fragmented* data itself, the personal energy ("ZU") that is "entangled up" in maintaining continuous creation and unknowing storage of *imprints,* also lowers *Awareness.* The "Human Condition" allows *Self* to *perceive reality* using the misinformation continuously fed to it through low-*Awareness* states. Such is the recipe for *hypnotism.* Thus, "standard-issue" Humans are suspended in a kind of *hypnotic state.*

There are many types of *imprints,* but those that contain the greatest "energy-mass" are the ones that cause the greatest *fragmentation.* In essence they "restrict"—or increase *resistance* against—free-flow of *Spiritual Life Energy* along the *ZU-Line.* They also "restrict" an individual's *Awareness*—and *perception* of *reality*—to rigidly fixed specific considerations rather than having a full-range to choose from and experience.

At an organic level, the cellular and genetic storage of *imprinting* once served an "evolutionary" purpose. Automated response-reaction mechanisms assisted in protecting organic life in the absence of any higher form of reasoning. In the 1960's, those involved in "consciousness expanding" "psychedelic experiments" stumbled onto this phenomenon and began referring to its data as "cellular memory" or "genetic memory."

The *Alpha-Spirit* is also capable of *creating* its own automated-mechanisms of the Mind-System. This happens *knowingly* at first; but, as the circuitry is *validated* and strengthened with increased usage, the "automation" be-

comes more solid. The individual eventually loses conscious *awareness* of the activity, yet *unknowingly* continues
to compulsively *create* these mechanisms with their own
*Spiritual Energy* ("ZU").

The Human Condition (the *Alpha-Spirit* + *genetic-vehicle*
combination) includes *fragmented* "thought" and "considerations"—but if that were *all* it entailed, it would be a
matter of simply "changing our minds" about *reality*,
&tc. And the truth is, it *should* be that easy. But another
layer of truth reveals that our *considerations* and *beliefs* or
*reality-agreements* are actually reinforced quite solidly in
*Beta-Existence* as the experience of "emotion."

Unlike "analytical thought" alone, the experience of
"emotions" includes a physiological response by the
"body" (*genetic-vehicle*). When maintaining a low-order
of *Awareness*, such *sensations* actually reinforce (or "*validate*") the *reality* of something—and these are *duplicated* as
*reality* in one's own "Personal Universe."

For example: the *Alpha-Spirit* is not actually located in
physical space-time or energy-mass bodies of *Beta-Existence*. We experience the Human Condition *remotely* from
*outside* this Universe—but we also *create* a *facsimile-du-
plicate* of the data *perceived* from the Human experience.
This is an *Alpha-Spirit's* "Personal Universe." It also includes any data and potential "*mental imagery*" that may
be *created* "mentally" at *Will*.

When we consciously (*knowingly*) use our *Awareness* to
*create* "*mental imagery*" (or confront any data) in the
"*Mind*," we *realize* our "ownership" of the *creation* and
have the ability to also "*uncreate*" it at *Will*. But, in the
case of "*imprints*" (*fragmentation*), an experience of the

*"imagery"* and its *triggered* emotional response is *unknowingly created*. Energy is compulsively suspended or entangled in maintaining its continual *creation*.

That which we are *freely* able to "consider" (or *create*) does not generally give us much trouble. But then, most of what we would choose to *create* is desirable or interesting, if not pleasurable. The contents of an *imprint* are usually things we *don't* want to "confront"—thus, we will not *perceive* or "see" a thing *"As-It-Is."*

*Imprints* are very much like "snapshots" of *perceived reality*. The experience of them as a *"mental image"* is not restricted only to "visual" scenery. *Imprints* include a record of all *facets* experienced in the *"imprinting incident."* By *facets*, we mean *all* data—yes, visual, but also, sound, smell, lighting, time of day, temperature, humidity, objects present, persons present, body positions, motions... All of this is recorded *and* associated.

The type of *"imagery"* (or content) that we are reactively unable or unwilling to *confront* (*"As-It-Is"*) is generally *considered* "dangerous" or "painful" or similar. Whatever it is, it is *perceived* to threaten optimum survival conditions—reinforced and inflated by an *emotional* "charge" and physiological ("body") response. This prompts "unpleasant" *feelings* when *confronted* with any *facet* previously *imprinted* as "unpleasant."

Yet, what we are unable to *confront* is what we are constantly being faced with or chased by. Data from the *imprinting incident* is continuously created and stored (and automatically retrieved) as part of our defense-mechanisms to help us avoid danger and pain. This once seemed like a good thing. But since this occurs automatically and obscures a clear view, it is *fragmentation*—inhibiting one's free considerations of thought.

## CHANNELS AND CIRCUITS

In *Systemology*, the *ZU-Line* of the *Standard Model* is a theoretical construct or graphic representation of the *Alpha-Spirit* as a unit of *Awareness*. It is, in itself, the primary "channel" of communication—or energy-flow—that is directed by the *Alpha-Spirit* as an *Awareness*; otherwise considered the "perspective" from *Self*.

As an *Awareness*, our "intended" focus on some "*thing*" establishes a present-time "channel" of communication (and energy) that is best defined as "*attention.*" Essentially, anything that we have our "*attention*" on, we are making "contact" with using a focused "energy beam" of our *Awareness*.

When we refer to a *thing*, we generally mean its *form* and *substance* as a "mass" or construct. We also share an "*affinity*" with all *things*—either an impulse toward or a repulsion from; an emotional response of some kind. Because any *thing* (even as a concept) is a potential contact point for our "channel" of communication (and *attention-energy*), we refer to any *form, object* or *person* as a "terminal" in our *Systemology* vocabulary.

The relationship we have with any *terminal* will be evident by the quality of *channel* that is experienced. In most cases, the nature of the "energy-flow" will be determined by previous experiences and/or *reality-agreements*. The idea that we "learn" from "experience" and store data as "memory" is not a new concept. The way in which it occurs and how it affects our *perceptions* is, however, rarely explored systematically.

Early in the existence of a personal *"Spiritual Timeline,"* the *Alpha-Spirit* starts out very much like a "clean slate"—engaging in *creation* and *existence* as a "being" with practically unlimited power and range of willingness to experience. We were once willing to extend *"attention"*—and therefore "reach"—anywhere.

As *fragmentation* accumulates (and *Actualized Awareness* declines), an individual is less likely to *be, know* about, *think* of, *act* toward or *own* various "*things.*" This is, again, evident in the quality of the "channel" maintained with any representative *terminal.* If turbulent, there is likely an *imprint* strongly influencing it. As *Awareness* declines, a being is willing to engage with fewer *terminals* and "reach" less into *existence.*

*Fragmentation* is not only composed of "emotional encoding." Even when it is, there are many factors that can make it seem quite "rational." Sometimes a *fragmented channel* is simply "cluttered with debris." It may concern what first appears like "factual data"—but is misaligned, misappropriated or just false altogether. Thus, *"defragmentation"* concerns "clearing the debris" (or "energy blocks") from *channels* to any *terminal.*

In *Systemology,* we recognize that a *channel* consists of three primary "circuits" that we use to *perceive* an experience with any *terminal.* These *circuits* are distinguished by the flow-type of *attention-energy* an *Alpha-Spirit* employs. For example: the first *circuit* is an "out-flow" of energy—meaning our experiences when "extending" or "reaching toward" a *terminal*; what *we* have done *to* or *about* some *terminal*.

The second *circuit* is the "in-flow" of energy we receive from the environment or *terminal*—or else what the *terminal* has done *to* or *about us*. And finally, the third *circuit* concerns "cross-flows" of energy that we are able to observe, which means what *others* have done *about* or *to* the *terminal*. As a form of "social" learning, this also includes what we have "overheard" others "communicating"to others about the *terminal* as well.

Data from these three *circuits* affects the general overall quality of the *channel* we use to experience *reality* regarding a certain *terminal*. Much like the *"emotionally encoded imprints,"* the information and *facets* stored from direct experience or observation can easily serve as a "filter" that screens present-time experiences. In other words, it *unknowingly* affects what we are "reaching" and "withdrawing" from in our everyday *life*.

The solution toward freeing up the original stores of energy we possess—and therefore widening our range for thought-consideration and experience—is quite similar to the philosophy behind other *Systemology techniques*. A *Seeker* examines their *unknowing* ("*not-known*") participation with, and agreements about, *reality* at an accessible "above-the-surface" level of *Knowingness* (and *Self-Determinism*). The act of *living deliberately* is a practiced skill in this world.

## LEVELS OF ACTUALIZED REALIZATION

*Systemology* does not ignore a healthy use of what most consider *"intuition"* — or else *actual* high-level *Knowingness*. Of course, many individuals misappropriate or misinterpret *"intuition"* with being the same as "how they *feel"* — and these are not the same thing.

True *intuition* is not a "conditioned" or "programmed" *Knowingness*; it comes from "higher faculties" and not "lower" *emotional responses* which may be easily misaligned with a few painful experiences — or even one big one. True *intuition* is unlocked by increasing our levels of *actualized realization* and *clear perception.*

*Self-Actualizing* the first level of understanding means being able to affirm in *Self-Honesty*: "I act as I will." This does not mean "doing whatever" without regard to consequence. On the contrary, a *Self-Actualized* individual, free of *fragmentation,* is able to make the best possible decisions in regards to *Life.* The *Seeker* is regaining control — no longer allowing the *genetic-vehicle* to simply run itself solely on "stimulus-response."

This first level concerns control of physical behaviors of a *genetic-vehicle* — primarily by maintaining *Self*-control of "emotional states" in *Self-Honesty.* Only then can we actually be *Self-determined* in our actions — and, of course, clear in our "efforts" on a physical level. This is practiced in basic *Systemology* exercises.

The second level of *Actualized Awareness* concerns the Mind-System directly: a *Self-Honest* handling of thought

—the realized freedom and ability to *"Think as I will."* We mean literally, the ability to "think clearly" and consider a full range of thoughts without inhibition or negative emotional responses. In *Systemology Processing*, we work toward this level of *Awareness* with *"Beta-Defragmentation"* techniques.

Higher levels of *Awareness* extend beyond the Mind-System and pertain to the *Alpha-Spirit* as a *Beingness.* A third level of potential *Awareness* concerns *"Creating as I will"* in an *"Alpha-Existence."* Beyond this, we stand at the *Gateways of Infinity* and can confirm *"I am as I will."* Studies and techniques focused on these higher levels are treated in *Systemology Processing* as *"Alpha-Defragmentation."*

Collectively, this systematic journey towards a "higher" and more ideal "spiritual" state of *Awareness* is referred to as *"The Pathway."* An understanding of this *"Pathway"* may be earned by studying our philosophy and applying its techniques. While this *Basic Course* series is intended to provide the principle *"Fundamentals"* that all *Systemologists* should know, there is still much ground ahead for a *Seeker* to explore.

## PRACTICE EXERCISES

1.  Look around the room. Spot an object that you
    like. Focus your attention solely on the object and
    nothing else; nothing else in the environment and
    no internal activity of the Mind. Do your best to
    simply *be* with the object, but also committing its
    "form" to memory—without analyzing any reac-
    tions or other thought associations. Now, close
    your eyes and "mentally" *imagine* a copy (or fac-
    simile) of the object—*imagining* its details as
    closely resembling the "original" as possible. Dis-
    solve it, collapse it, or throw it away—then make
    another copy. Do this several times until you are
    satisfied with the appearance of the copy and the
    *Awareness* that you are knowingly creating the
    *mental image.* Then, intend for this copy to be
    "brighter" than before, and see it appear as such;
    then "dimmer." Practice this alternation a few
    times. Now, intend for its apparent color to
    change: make it "blue"; change it to "red"; turn it
    "green"; then finally return it to its original color.
    Dissolve this image and then open your eyes. Re-
    cord any *realizations* you may have had in your
    journal.

2.  Recall a time something seemed *real* to you. [It
    could be an object, a place, a person, or an event
    (or situation) that contains any of these—which
    considered *facets* of the incident.] Take a moment
    to notice details about the *mental imagery* that is

"called to Mind." Are there parts that immediately come into view? Are there parts you have to try hard to *imagine* or "fill in"? Practice this with various *"things"* (*terminals*). After spending some time with this, select one of the times recalled, then recreate it, but *intentionally* changing some of the data viewed. You might change the color of an object, or replace a person present with another individual, *&tc.*—whatever it takes to demonstrate your ability to *knowingly alter* the recalled data.

3. Get the sense of making the "body" sit (or lie down). Focus all your *Awareness* (*attention*) on just the feet, without considering the remainder of the legs or the rest of the body. If this proves too broad of a scope at first, focus on just one toe of one foot. While concentrating *Awareness* in that location, *imagine* the feet as nonexistent—or else, *consider* that if the feet *were* nonexistent, then *Self* would still continue its (*Alpha*) existence, unchanged, as an *Alpha-Spirit*. Then, *consider* that feet are useful for moving the "body" (*genetic-vehicle*) across "space" (in *Beta-Existence*), but that they are *not* the actual "feet" of the *Alpha-Spirit*; and the true *Self* is in no way dependent on physical "feet" in order to "act." It is important for an *Alpha-Spirit* to be in "good communication" with the *genetic-vehicle* to get along well in *Beta-Existence*, but it should in no way confuse its own *identity* with that of a *body*.

4. Refer to the previous exercise. Continue focusing *Awareness* on the other remaining points of the

85

"body"—the legs, pelvic region, stomach, chest, arms, hands, neck, and head. Treat each point with the same *considerations*—and move off from each with the same *realizations*—as with the feet. When these points have all been spotted, the *Seeker* may *consider* the whole body of the *genetic-vehicle* as a useful instrument for communication and activity in the *Physical Universe*. It is biologically adapted for organic life in *Beta-Existence*—but *Self*, the "I-AM" or *Alpha-Spirit*, *is* above and superior to, independent and apart from, the *genetic-vehicle*. The ideal "end-point" of this exercise is to get a sense of *Self* actually operating from a *Spiritual "Alpha"* point of *Beingness, exterior* to a *genetic-vehicle* and *Beta-Existence* altogether; but this *realization* cannot be forced prematurely.

5. Make two lists of significant *"terminals"* in your life: "things" you like very much; and "things" you don't like at all. Remember: *"terminals"* are "things" consisting of "mass"—persons, places, objects, *&tc*. Use several sheets/pages as needed, but keep the two lists separated. Of course there are many things encountered in *Life*, but we are interested in the most significant ones that readily "come to mind." When the lists are satisfactorily completed, take each "item" listed and "scan" your surface memory for instances or occurrences that involved that *terminal*. Alternate between the two lists to avoid excessively focusing on the "negatives." See if you can get a sense or idea of *why* you "feel" a certain way about a particular "item." It may help to refer to the sect-

ion on "Channels and Circuits" to target specific "types" of energy-flow that pertain to a *terminal*. If you find that "thinking about" an incident involving a "thing you don't like" is causing you major discomfort: alternate *spotting* something in the incident and *spotting* something in the room (or environment) until its handled.

6. Using your lists from the previous exercise: take each listed "item" and *practice* intending your "feelings" about that particular *terminal* to change to the opposing list. Once you have a sense of it, intend your "feelings" to switch back to how they were. Practice alternating your "feelings" in this fashion. It may be easier to treat items you are more "neutral" on at first. The exercise is simply a demonstration of how the control of such *considerations* ultimately rests with our *Self*. Obviously the goal is not to get a *Seeker* to dislike things they like, but to establish just how "fluid" our feelings actually are. Ideally, we would like to see, at the very least, a "softening" of the harsher emotional responses that "things we don't like" seem to inspire in us. It does not mean we have to "like" such things or engage with them; but, we want to reach an acceptable tolerance level that enables us to *confront* (or "face up to") their *reality*—to be at "*cause*" over their handling and how they are duplicated (or "copied") in our Personal Universe, rather than reactively withdraw and unwillingly (or *unknowingly*) remain an "*effect*."

7. Seat the "body" comfortably in a quiet room.

With eyes closed, "mentally" (*intentionally*) "reach" out with your *Awareness* and hold all of your *attention* on the two rear "upper-corners" of the room (the corners *behind* where the body is located). *Intend* to focus all of your present-time "*interest*" on only these "corners" and think of nothing else. That is the totality of the exercise. If other "thoughts" invade your *Awareness*, simply return all of your "*interest*" on the "corners." Hold this *intention* for as long as you can, working up to handling longer durations of time with practice. An hour or more of actual practice is effectively more beneficial than any other known traditional "meditation" technique. "Corners" of a room are "points" that define the "*space.*" A room has eight of these "anchor points" in all. This exercise may be extended to gradually increase the number of "corners" that you can hold your *attention* on simultaneously.

# LESSON FOUR:
# ANCIENT
# SYSTEMOLOGY

## LESSON FOUR
## WISDOM FROM THE ARCANE TABLETS

After studying previous lessons (booklets), a *Seeker* may get to wondering about our interpretation of data drawn from the *Arcane Tablets*; wondering if our presentation of various *models* and *charts* is truly a stable foundation of fundamentals to base our "*Systemology of Everything*" on —if it is something a *Seeker* or student can put their *trust* in as something "solid."

Our philosophy is not one based on *faith* or *dogma*—it is rooted in a *systematized* understanding of some of the most ancient writings found on the planet. While other civilizations may have come and gone in the distant past, these *esoteric* and *hermetic* teachings are those which the "highest minds" based the upper-level systems behind *this* current incarnation of civilization on many thousands of years ago. If anything, it is *this* information that has actually survived from the even more distant and long-forgotten past. And a perfected communication of it has not yet occurred in modern times prior to our *Systemology*.

The focus of this lesson (booklet)—unlike the previous ones—is not on the explanation of specific topics or subjects. With the previous lessons in mind, or on hand as reference, we take a step further here by delivering the fundamental "*axioms*" (basic principles) of the *Arcane Tablets*—the "raw" data underlying our original research and interpretive models.

Therefore, what follows below are the *essentials* of "Ancient Systemology" that we've carried over into this "New Age." Our knowledge base is otherwise unparalleled in modern *metaphysics* and *esoterica*. We have modernized the language used to now communicate the ancient wisdom set down thousands of years ago.

Much of this data is self-explanatory for a continuing student—providing much for a *Seeker* to study or *reflect* on in view of previous lessons. Individual elements are explained more thoroughly in other lessons, and in the text: *"The Tablets of Destiny Revelation."* In essence, this is the raw data on which all the former lessons are based. This lesson also collects the sum of this former instruction into a workable *Systemology*.

---

## THE TABLETS OF DESTINY

All data and information is understood as *knowledge* only to the extent or level of *Awareness* that it is *processed*.

*"Understanding"* and *Awareness* have a tendency to "rise and fall" together, building upon one another to form levels on which we base our experience of *Reality*.

The three tiers or levels of processing or understanding include: what is *physically* observed; what is *intellectually* realized; and what is *spiritually* applied.

There are two primary types of knowledge: the *"exoteric"* understanding that is held by the common masses, species or group; and the *"esoteric"* understanding that is restricted to a certain actualized level of *Awareness*.

*Self-determinism*—the ability to *Self-direct* creation and/or enact change or action—is proportional to one's level of understanding and responsibility.

The ALL ("AN-KI") is the latent unmanifest potential of *Everythingness*.

The LAW—represented on tablets as a "cosmic dragon"—is an existential division between the *Spiritual Universe* ("AN") and the *Physical Universe* ("KI").

This *Cosmic Law* or *Cosmic Ordering* defines the interacting systems of manifestation present in the *Universe*—*consciousness*, *motion* and *substance*. It consists of the *reality-agreements* that define the *Physical Universe*.

From an *Alpha* state of existence, *Self* engages with an "organic body" to experience a *Beta* state of existence. *Self* is a "spiritual cause" of "physical effects"—engaging a *Self-Determined "Will"* as *Actualized Awareness*.

The "spiritual energy" transmitted to all *Life* (as *"Life-force"*) goes by many esoteric names throughout history —but we find the concept first treated as "ZU" on cuneiform tablets from Mesopotamia.

A "conduit channel" of *Spiritual Life Energy* ("ZU") links *Awareness*-levels of our *"I-AM-Self"* from the *Spiritual Universe* (*Alpha*) to the degrees of variation experienced as *"effects"* in the *Physical Universe*.

The term *"levels"* applies best to the relative tiers of personal *Awareness* and understanding regarding *Reality*. These *"levels"* are reflected in the steps of the "ziggurat" temples in ancient Mesopotamia—the original *"Stairway to Heaven."*

The term *"degrees"* is best applied to the *variation* in form and activity perceived by a *"Beta"* state of *Awareness* interior to the *Physical Universe.* This includes "emotional" and "mental" states inherent of the *"Human Condition."*

The *"Self"* is "I" as *"Spirit"* regardless of the position it considers its perspective to be "looking" or "feeling" from. [This true *Alpha* state of *Self* is what some call the "spirit" or "soul."] Both, its *Beingness* and the "conduit channel" of *Spiritual Life Energy* ("ZU") originate from an *Alpha-Existence*, separate and apart from the *Physical Universe.*

It is on this *Lifeline* of "ZU" energy that the *Alpha-Spirit*—operating from the *Spiritual Universe* "exterior" to *Beta-Existence*—is capable of *Self-directing* (*intending*) "thought" into "action" at various degrees of manifestation in the *Physical Universe* (*Beta-Existence*).

---

### ORIGINS OF THE STANDARD MODEL

In *Systemology*, we refer to the continuum or "spiritual conduit" of "ZU" as the *"ZU-Line."* Using this idea, we may graphically represent "activity" of *Awareness* along an entire "spectrum" of potential ZU "states"—an activity that is generally referred to simply as *"consciousness"* in many other sources and traditions.

When sources mention "states of consciousness" they are referring to various "gradients" of vibration (or frequency) where *Awareness* may be fixed along the *ZU-Line.*

Each "point" or "degree" on the *ZU-Line* has a certain vibration or frequency that defines its "quality." [To establish a "Standard Model" we assigned a relative "numeric quantity" to these *degrees* in order to figure a logic of comparison to other *degrees*. Where it pertains to the "Human Condition" we refer to it as the *Beta-Awareness Scale* in "Lesson 1."]

Graphic representation of frequencies on the *ZU-Line* may include such *degrees* as: a "zero-point" organic *body* ("genetic-vehicle") death; cellular activity and sensory perceptions of the *body*; chemical production induced by *emotion*; *thought* vibrations transmitted between the *Alpha-Spirit* and "genetic-vehicle."

There are also points on the *ZU-Line* "exterior" to *Beta-Existence*, where the *Alpha-Spirit* "imagines" and "intends" (uses *Will*). These are *Alpha* qualities that originate from outside of the *Physical Universe*.

We may experience the perspective of any point or *degree* along the entire continuum (*ZU-Line*) from *Self* as the *Alpha-Spirit*. But, the "I-of-the-Observer" remains *Self* in this *Alpha* state, regardless of where *Self* considers its "point-of-view."

Each of us is *knowingly* "projecting" the totality of our *Spiritual Awareness* from an *Alpha* point that is *exterior* to the *Physical Universe*. This "projection" of *Awareness* remains in totality along the *ZU-Line* only to the extent that the "conduit-channel" for *Spiritual Life Awareness* is "clear" of "debris" (*fragmentation*).

*Fragmentation* is anything which inhibits the total experience of *Self* as "Actualized Awareness." The "Pathway to Self-Honesty" in *Systemology* is a personal journey of *clearing* the spiritual-energy *channel*—or *ZU-Line*—of the "debris" *imprinted* upon us by our environments and experiences.

## THE PATHWAY TO SELF-HONESTY

*Self* does not actualize an *Awareness* past a point not understood; our *Awareness* does not surpass our understanding, but it may support a reach toward a higher level of understanding.

When we *defragment* the knowledge-base of our understanding, we remove "dissonance" or else clear the old static beliefs that stand as solid obstacles *fragmenting* the *Awareness* of *Self*.

The *process* of *defragmenting Awareness* (on the "*Pathway to Self-Honesty*") involves closely examining the varying degrees of *fragmented* "thought" and "emotion" fluctuation along the *ZU-Line*.

There are three primary systems that communicate *Life* energy along the *ZU-Line*: spiritual systems, psychological/ mind systems and physical systems. These three systems also correlate with the principle systems of manifestation: consciousness, motion, and matter; and also the states of *Self-determinism*: being, doing, and having.

The "peak" level of *Awareness* that we chronically maintain is proportional to the frequency that our attentions and thought-energy is fixed to.

An understanding derived from the *Standard Model* and *ZU-Line* is the basis for "processing" techniques used in *Systemology*—and the map by which we chart our ascent on the "*Pathway to Self-Honesty*."

A holistic examination of *Arcane Tablet* lore revealed two

primary communication relay-centers along the *ZU-Line* —both pertaining to the "Mind-System"—acting as an intermediary between the *Alpha-Spirit* and the *genetic-vehicle*. They are not named directly on the original sources, so we titled them for convenience.

The primary *Master-Control-Center* (*MCC*) ["4" on the *Standard Model*] is the direct point-of-contact between the *Alpha* ("spiritual") systems and the Mind-System. It is a perfect computing device to the extent of the information received (and perceived) from "lower levels" of sensory experience.

A secondary *Reactive-Control-Center* ["2" on the *Standard Model*] exists lower on the *ZU-Line*, responsible for discharging emotional *biochemicals* and motor *impulses* of the organic body. It is also quite "impressionable" and stores emotionally-charged information in a form we call "*imprints.*"

The *degrees* experienced between "4" and "2" (under the *MCC*) pertain to "states" of thought-involvement with *Beta-Existence*. Those between "2" and "0" (under the *RCC*) pertain to "states" of emotional response.

All sensory input is *perceived* by the Mind-System and communicated back to *Self*. *Awareness* is communicated along this network of energy relays (along the *ZU-Line*) and forms a "feedback loop" with the *Alpha-Spirit* (*Self*). This activity is what some refer to as "*consciousness.*"

If *no* obstruction, distorted "*lens,*" emotional-entanglement or thought-formed solid is *fragmenting* lower-levels of energetic interaction (on the *ZU-Line*), than the actualized *Awareness* an individual experiences is in a condition that we call "*Self-Honesty.*"

## SYSTEMS OF THE HUMAN CONDITION

Experience of *Beta-Existence* mainly pertains to our "contact" with energy vibrations carried by everything and everyone we meet. That which is particularly significant concerns communications by family, friends, authority figures, those we admire, and others in our civic communities.

"*Reality*" is an agreement concerning "what *is*" in regards to thought, matter, and the considerations that define our experience. Our experience of *Reality* is either *Self-Determined* or under an "outside" control.

All "systems" are *dynamic* ("variable" and often "interconnected" to other systems) and subject to our participation for us to experience their qualities. The level of this "participation" is subject to our level of understanding and *Awareness*. This is what *is* in our ability to directly control in this lifetime.

Mechanisms of control for "*Reality*" are the same as the control *Self* maintains with other "systems." Short of the *Alpha-Spirit* "creating" a *new* "system," an individual is really only able to *start, stop* or *change* some variable of an existing system—or one system in a network of interconnected systems—such as in the operation of a "vehicle."

The *Standard Model* demonstrates that our thoughts are creating vibrations of varying degrees or frequencies "up and down" the *ZU-Line*. These personal vibrations interact with varying levels or conditions of the "world around us"—and even the "organic body" used to receive environmental stimuli.

Solidified thought-forms and *"Reality-Agreements"* that are intensively broadcast have the tendency to be carried and built upon by others who are affected by them or that in some way "share" in their level of *Reality*.

Even if a person is "opposed to" or "protesting" an *idea*, there is still a "thing" in *Reality*. There is now a "thing" existing for us to *agree* to be in any position about at all. This is an inherent part of "shared" *Reality*.

The *"interior"* (*Beta*) and *"exterior"* (*Alpha*) composition of an individual's *Reality* is treated along the same *ZU-Line* as *Awareness*.

The relay centers of the Mind-System do not perceive a difference between what is generated as a *Reality* "internally" within itself from what is sensed as the *"external"* world—nor does *Self*.

We continuously reinforce the *solidity* of our "thought-forms," "imprints," and "fragments," as *Reality* every time we *reactively* or *unknowingly* revisit them with new energy (*attention* or *interest*) and engage an automatic response (that is not *Self-Determined*).

Potentially harmful *"reactive-response"* mechanisms and *"emotional encoding"* may stay dormant in the Mind-System for long periods of time without *resurfacing* (being "restimulated") directly.

Painful memories ("physical" or "emotional"), biological unconsciousness, and other forms of physical trauma, are forms of *fragmentation* that result from having *Awareness* "suspended" for long periods of time (or severe instances) below "2" on the *Standard Model* (*ZU-Line*). These emotionally encoded *fragmented facets* accumulate as an "energetic-mass"—or *"Imprint."*

*"Imprints"* are fixed rigid energy-masses stored (at the *"RCC"* level of the Mind-System) as entangled emotional turbulence, operating *"below-the-surface"* of conscious analytical thought.

Using modern *"Systemology Processing"* techniques, a *Seeker* systematically *"resurfaces"* and *"confronts"* the content of the *Imprints*—thereby freeing up entangled masses of *Spiritual Life Energy* and increasing actualized *Awareness.*

## ARCHAIC AXIOMS

The following *"archaic axioms"* are derived from 20th Century "New Thought" interpretations of the *Arcane Tablets* that existed prior to our *Systemology*—and on which our earliest researches were grounded until establishing our own *"Fundamental"* understanding of the *Standard Model* and *ZU-Line.*

The Law *is.* Other than The Law, there is but *Infinity*— which is *Nothingness.* But in that *Infinity of Nothingness*, there is the unmanifest, the latency, the potentiality, the promise of *Manifest Everythingness.* And the *Nothingness* is counterbalanced by the *Everythingness* of the *Cosmos.*

Under the Law, the *Cosmos* is governed. Each and every thing, and all things, proceed in an "orderly trend"— which is to say "sequentially" and "systematically."

What humans call *"matter"* is the countless centers produced by *Will-Intention* in the *substance principle (system)*, through the action of the *motion principle.*

What is called *"force-energy"* is the action of the *motion principle (system)* upon the *substance principle (system)*, induced by *Will-Intention*.

*"Thought"* is the action of *Will-Intention* upon the *consciousness principle (system)*, employing the *substance* and *motion principles* in the operation.

When an individual attains *Self-Actualization* they enter upon the plane (*level*) of *Will-Intention*, and rise up above the plane (*level*) of *"Desire"* (*emotion*).

*Will-Intention* and *Desire* are "opposing poles" of the same principle (*ZU-Line*), the center or balance of which is *Reason*.

Operating from the plane of *Will-Intention*, one learns to use the *Law* to maintain being at *"Cause"* rather than being passively beneath it as *"Effect."* They may still *create* *"Desire"* by *Will*—or else, *Will* to *Desire*; but *Will* is no longer being diminished or influenced by *"Desire."* Above this, the individual can learn to *Will* to *Will*.

## THE SEVEN COSMIC LAWS

The *Arcane Tablets* reveal that the *Cosmos* is regulated by a "Cosmic Law"—actually *Seven Cosmic Laws*—superimposed over the *Universe*. It constitutes the most basic level of *reality-agreement* concerning experience of manifestation at the level of *Beta-Existence*. An understanding of "Cosmic Law" is inherent in ancient *"Hermetic philosophy"* in addition to *"New Thought"* teachings predating our *Systemology*.

I. *The Law of Orderly Trend.*

"Under this law, there is always manifested law and order in the Cosmos, from suns to atoms; from the highest to the lowest; matter, energy and consciousness."

II. *The Law of Analogy.*

"Under this law, there is found a correspondence and agreement between all of the various forms of manifestation. What is true of the atom, is true of the sun. What is true of matter, is true of energy and mind. To know one is to know all."

III. *The Law of Sequence.*

"Under this law, there is included the activities of what is generally known as 'cause and effect'. Nothing in existence happens by pure chance. Nothing happens without a precedent manifestation, and a subsequent manifestation. Nothing stands alone in exclusion."

IV. *The Law of Rhythm.*

"Under this law falls a variety of phenomena, the most important of which is 'vibration'. Everything in existence is in constant vibration—everything material, mental, or of 'energy'. Upon this fact depends the variety, degrees, states, and conditions, of the manifestations in the Cosmos. To control vibration is to control all forces in the Universe."

V. *The Law of Balance.*

"Under this law, there is to be found an explanation for the universal equilibrium, compensation, and balance, observed in all manifestation in the Cosmos. One thing balances another; everything has something set opposite it, to balance it."

VI. *The Law of Cyclicity.*

"Under this law is found the cyclic—or circular—trend of all things, physical, mental and spiritual. Everything moves in circular systems. The wise convert the circles into upward spirals. Instead of traveling and endless circle, or downward, the wise rise in spirals to attainment and advancement."

VII. *The Law of Opposites.*

"Under this law is to be found the explanation of the fact that everything has its opposite; everything *is* and *is-not* at the same time; everything has its *other side*—also the fact that opposite things are alike, in the end, for the extremes meet and contradiction may be reconciled."

## PRACTICE EXERCISES

1.  Get the sense of making the "body" sit (or lie down) in a comfortable position and in a quiet room (or uninterrupted outdoor environment). Close your eyes and using imagination, *"be outside the body."* Simply shift Awareness to a "point-of-view" just outside the physical body. Using your "exterior" Awareness, gaze upon its form. Even as imaginative practice, this exercise should promote a *realization* that the actual *Awareness* (and true "spiritual" existence) of *Self* is "exterior to" and separate from a body (or genetic-vehicle). Once you have worked with this a while: if you have objection, difficulty, or no reality on this practice, go to *Exercise #2*; otherwise, go to *Exercise #3*.

2.  Certainly a person could object that this is merely *"imaginary"* play—and since one can, in theory, *"imagine anything"* that it really "proves" nothing. Such a *Seeker* might better consider this another way, as the old esoteric instructors once relayed—in trying to *imagine* yourself as "dead." In this wise, all that is generally accessible is the concept of a dead physical body; meanwhile, the actual *Awareness* (of *Self*, the *Alpha-Spirit*) might either "stand" apart from the "dead body" (and be able to view it), or else remain behind to inhabit a "dead body." In either case, what is generally considered *"consciousness"* itself, would not

perish or cease at its own level of existence simply because of the "organic death" of a *genetic-vehicle*.

3.  This practice is students that have achieved some sense of *Reality* on the first or second exercise. Focus your *Awareness* (*attention*) on a singular "spot" (or "point") internal to the physical body. You may use your *imagination* to get a "sense" of this "spot" (or "point")—since you won't be able to use the body's "eyes" for that part. Then, focus your *Awareness* (*attention*) on a singular "spot" external to the physical body, in the room (or immediate environment nearby). Get real interested in that "spot" for a moment. Now, fix your attention on the "spot" inside the body; then alternate attention to the "spot" in the room. Once you are proficient in practicing this exercise, continue below.

4.  Using the previous exercise as a guideline, practice alternating *Awareness* (*attention*) between "three spots in the body" and "three spots in the room." This may be practiced for two minutes or two hours. The body's "eyes" may be open or closed.

5.  Using the previous exercises as a guideline, practice alternating *Awareness* (*attention*) as before, but with *eyes closed*. If necessary, use your *imagination* to get a "sense" of the "spot" in the room—since you now won't be able to use the body's "eyes" for that part either. Of course, practices like these, even if only *imagined*, may "turn on" actual *perception* as if viewing from a point re

mote from the body. Don't worry if this does not happen immediately—or even at all—during your first practices.

6. When you are comfortable with the exercises above, practice alternating between "spots in the room" and "spots outside the building." You can begin with eyes open for "spots in the room"—but you will obviously be employing or *imagining* "spirit vision" (or a "remote viewpoint") when looking at "spots outside the building" (or, if outside, looking at "spots" in the surrounding environment that are not visible with the body's "eyes"). At first, just get a "sense" of looking at the "spots" external to the building, just as you did with "spots" internal to the body. With practice, you may find it possible to perceive actual environmental data in this wise.

# LESSON FIVE:
# EVOLUTION OF A
# SPIRITUAL SCIENCE

## LESSON FIVE
## EVOLUTION OF A SPIRITUAL SCIENCE

In these *Basic Course* lessons, we are not only concerned with communicating what the "Foundations of Systemology" are, but also in establishing a solid foundation of understanding for the *Seeker* or student that is pursuing these studies. A *Seeker* is not likely to "apply" a philosophy that isn't solidly "*real*" to them.

In "*Lesson 4*" we explored the basic "raw" philosophical data directly derived from ancient *Arcane Tablets*. In this lesson (booklet), we further solidify the foundations of our philosophy by highlighting just a few of the more modern sources of inspiration that led to the evolving development of *Mardukite Systemology*.

This *Basic Course* lesson is a rather incomplete rendering and imperfect tribute. It does, however, serve our purposes for introducing a variety of related figures and subjects. It demonstrates that we did not generate the idea of *Systemology* from nothing. But, this is *not* a student's "required reading" list, or even necessarily a "recommended reading" list. It is simply a supportive *bibliography* of sorts—as our work seldom has one.

Our *Systemology* is certainly not based on any *one* of these sources exclusively; nor was *any* one single source taken at "face value" all inclusively during our research phases. And finally, the total culmination of all these points and facets, if simply summed up together, does not equal *Systemology* either. Yet, we did not develop our philosophy in a vacuum; and *Systemology* is a predictable evolution of former understandings.

*"There is a correspondence between
things seen and unseen;
earth is the shadow of heaven and
humanity is a reflection of divinity."*
—Eliphas Levi


*"Make use of life, its course is so soon run;
yet 'systemology' teaches you how time is won.
I counsel you, dear friend, in sum, that first you
learn 'collegium logicum'.
Then the best of all worth mention—
to 'metaphysics' you must pay attention,
And see that you profoundly strive to gain
What is not suited for the human brain."*
—Goethe, *'Faust*


*"From the senses arises opinion; and from reason,
demonstration. On the former are huddled up the
prejudices of the vulgar, following the bare
'exoteric' appearance of things; on the latter are
founded the 'esoteric' axioms of the wise, who
consider things as they are in themselves."*
—John Toland, *'Clidophorus'*

Unlike other installments of the *Basic Course*, this portion does not include "practice exercises" pertaining to its study. As an alternative, a *Seeker* or student is encouraged to personally reflect on the "quotations" cited throughout—and also independently research any aspects of this brief history lesson that may require additional verification for an individual's own intellectual satisfaction.

---

## JOHN TOLAND 1670 – 1722

John Toland, a notable figure of "free-thought" during the *Age of Enlightenment*, is best known for sparking the first modern European "Druid Revival" in 1717. Rather than an emphasis on Celtic mythology, Toland's Druids are "high-minded" intellectual philosophers and scientists, not dissimilar from "Freemasons" that officially emerge in England the same year and even met at the same location there: *"The Apple Tree Tavern."*

Toland's work *"History of the Celtic Religion and Learning: An Account of the Druids"* appeared posthumously in 1726, compiled from personal correspondence during his lifetime and prompting *Antiquarian Druidism*. In 2018, the present author published an annotated tercentenery edition of Toland's *"Pantheisticon"* for existing *Systemology Society* members—a short text on natural philosophy and esoteric-philosophy societies.

Key contributions include: a distinction between *"exoteric"* philosophy—what is commonly understood and discussed in public—and *"esoteric"* philosophy, which is communicated privately by a few; and first coining the term *"pantheism"* to describe a philosophy of "divine energy" imbuing all *things,* and a *"Cosmic Law"* that dictates the qualities, patterns or limits for the "energetic interactions" of *things* with each other.

## HELENA P. BLAVATSKY AND
## THE THEOSOPHICAL SOCIETY

Helena Blavatsky (1831–1891), a spiritualist, co-founded the "Theosophical Society" (*New York City*) in 1875, sparking an international movement that widely incorporated Hindu and Buddhist concepts into Western mysticism. Her classic works—*"Isis Unveiled: A Master-Key to the Mysteries of Ancient and Modern Science"* (1877) and *"Secret Doctrine: The Synthesis of Science, Religion and Philosophy"* (1888)—possibly demonstrate the best 19th Century attempts of systematizing "ancient wisdom."

It is perhaps the *Theosophical* movement as a whole—rather than Blavatsky individually—that influenced the present author's early researches into what would eventually develop into *Systemology.* The movement parallels *"New Thought"* in many ways—except that it mostly attracted philosophical-types with mystical ambitions for *Ascension*, rather than targeting professionals seeking luck in business as *New Thought* did. *Theosophy* also reintroduced the disputed scientific concept of *"ether,"* but referred to it as *"Akasha."*

The *Theosophical Society* provided a forum that many creative spiritualists contributed to, including C. W. Leadbeater (1854 –1934), Jiddu Krishnamurti (1895–1986), Rudolf Steiner (1861–1925)—and particularly, J. J. van der

Leeuw (1893–1934), whose *"Conquest of Illusion"* (1928) is quoted from directly in our text: *"Imaginomicon."*

Key concepts include: modernization of "Eastern" mysticism, the "chakras" and the "cabala"; physical and non-physical systems as interrelated; personal, interplanetary and cosmic evolution; the "Observer-effect" on *Reality*; acknowledgment of an unseen (metaphysical) "council" or "brotherhood" of Ascended Masters that have directly initiated (or communicated with) select individuals on a "path" toward *Ascension*.

## EDWIN ABBOT ABBOT
### 1838 – 1926

Written over a century and a half ago by a university headmaster, *"Flatland: A Romance of Many Dimensions"* (1884) remains a timeless relevant classic of philosophy, mathematics, and dimensional-theory. It is a fantasy narrative—comparable to *"Alice in Wonderland"*—told to us by the character 'A. Square' who lives on *Flatland*, a two-dimensional plane inhabited by sentient living 'shapes'.

The "model" that Abbot uses for this fantasy are easily demonstrable to us: the interaction of a *"third-dimension"* with a *"two-dimensional"* plane (or *Flatland*). The "higher" *third-dimension* is not perceived directly by inhabitants restricted to *two-dimensions* of sensory awareness.

*"A. Square"* has an encounter with a sentient unseen "sphere" that is able to speak and interact from the *third-dimension*. The comparison in the narrative rings true regarding limitations of our perception of a "higher" *Univ-*

*erse* or *dimension* while restricted to only perceiving through the human sensory range.

Contribution: *"Flatland"* illustrates inter-dimensional metaphysics quite clearly, introducing an advanced concept to everyday readers in plain language. As such, it ranks fairly high on the present author's "suggested reading" list—particularly for *Seekers* having a difficult time conceiving of a metaphysical *Spiritual* (*"Alpha"*) plane interacting with *Physical* (*"Beta"*) existence. This interaction was once called "unseen perturbation" in *archaic systemology.*

---

## WILLIAM WALKER ATIKINSON
### 1862 – 1932

Atkinson was a pioneer of the *American "New Thought"* movement. While editor of *"Suggestion"* magazine (later *"New Thought"* and *"Advanced Thought"*) in 1900, he began writing books—leading to the founding of the "Yogi Publication Society" (in *Chicago, IL*)—and then served as president of the *"International New Thought Alliance"* (which still exists today). He eventually wrote over 100 books in 30 years under various names.

By 1920, his publishing company produced dozens of titles for their catalog—with Atkinson using different pseudonyms (such as *Yogi Ramacharaka* and *Theron Q. Dumont*) when writing about specific traditions. His anonymous works—*"Secret Doctrines of the Rosicrucians"* (by *Magus Incognito*) and *"The Kybalion"* (by *Three Initiates*)—are quite famous and widely-circulated among "New Age" *esoteric* researchers and practitioners.

"In phenomenal existence there is nothing that is
 independent of everything else.
Everything is a 'degree' or 'aspect' of whatever
 Everything is.
Our expressions are in terms of Continuity.
If all things merge away into one another, or
 transmute into one another, so that nothing can
 be defined, they are a oneness, which may be
 the oneness of one existence.
It is not exclusively anywhere where anything is,
 if ours is one organic existence, in which all
 things are continuous."

—Charles Fort, 'Book of the Damned'

"It is 'you' who lives on forever, not some
 intangible thing or soul that develops from you
 at the hour of death.
This 'you' is living in Eternity as much now as it
 ever will be.
This is Eternity—right now. Many of us, before
 we grow into an understanding of things, feel
 that this life is of no consequence—that it is a
 miserable thing and that true living will not
 begin until we get out of the body and
 'become' a Spirit.
You are a Spirit as much now as ever."

—William Walker Atkinson

Atkinson's "New Thought" was a westernized *American* "answer" to European occultism, Theosophy and other new popularly rising interests in "Eastern" mysticism. It surpassed considerations of the former "transcendental" movements—providing the first truly *American* tradition of *applying* esoteric philosophies.

*"The Law of the New Thought: A Study of Fundamental Principles & Their Application"* (1902) and the numerous volumes comprising his library of *"Arcane Teachings,"* proved essential to the founding of *Systemology* as an "applied philosophy." In addition to being a source of significant inspiration for the present author since the 1990's, Atkinson's works are respectfully quoted from several times in our own *Systemology* literature.

Key concepts include: *Self* operating independent of a body; *Self* as "being soul" rather than "having a soul"; *Self* having *one* "spiritual life" but experiencing *many* forms or "lives"; *Self-Actualization* as a journey of clearing fragmentation from many *"consciousness*-levels" for personal *Ascension*; and that activities of *this* lifetime, either toward increased *realizations* or toward degradation, carry into experiences of future *lives*.

## CHARLES FORT 1874 – 1932

Fort was a self-taught scholar, researcher, and journalist, residing primarily in New York at the turn of the century. His interest in "curiosities" and "anomalies" of natural phenomenon led to the first and most widely-circulated of his works, *"The Book of the Damned"* (1919).

*"Book of the Damned"* is titled for the idea that such know-

ledge remains outside the current general understanding, which science generally rejects or ignores, and is therefore "damned" or "excluded." The book pioneered the new field of "anomalistics" and inspired a cult following of "*Fortean*" philosophers. It also includes the first in-depth exploration into UFO-phenomenon (including triangular patterns) decades before common sightings began in the 1940's.

Key concepts: "monistic continuity" (the *ZU-Line* is a "monistic continuum" a "*singularity force*" expressed across an infinite line demonstrating all possible degrees of interaction or manifestation); and "unseen perturbation" (visible phenomena instigated, triggered, or "*perturbed*," by unseen variables or influences).

---

## CARL G. JUNG 1875 – 1961

Carl Jung—originally a serious student of Freud's "*psychoanalysis*"—is most famous for developing "analytical psychology." It is perhaps the best and last attempt to actually understand the "Mind" and "Spirit" of an individual from within the academic confines of "psychology" as a medical field.

Jung's work, while clinical, incorporated mystical themes such as alchemy, Druidism, intuition, past-lives, and even UFOs. The therapeutic and pop-cultural practice of analytically "typing" an individual's *personality*—such as "introversion" and "extroversion"—is also mostly derived from *Jungian Psychology*.

As an impressionable youth in the 1990's, Jung's books—and books by those sharing his brand—inspired the

*"We should know what our convictions are
and stand for them.
Upon one's own philosophy, conscious or
unconscious, depends one's ultimate
interpretation of the facts.
Therefore, it is wise to be as clear as possible
about one's subjective principles.
As the man is; so will be his ultimate truth."*
—Carl G. Jung

*"My life is a story of the self-realization
of the unconscious.
Everything in the unconscious seeks outward
manifestation, and the personality too desires
to evolve out of its unconscious conditions
and to experience itself as a whole.
I cannot employ the language of science to
trace this process of growth in myself, for I
cannot experience myself as a scientific problem."*
—Carl G. Jung

*"Human perception involves 'coding' even more
than crude 'sensing'—thought is abstraction...
Language, mathematics, the schools of art, or any
system of human abstraction, gives to our mental
constructs the structure, not of the original fact,
but the symbol-system into which it is coded."*
—Robert Anton Wilson

present author's interest in pursuing "psychology" (as a means of bridging the understanding between "magic" and "science"). But, after a complete disillusionment with the field at a university level, the author left to independently pursue any related research that might later contribute to establishing *Systemology*.

Key concepts include: existence of universal "archetypes"; potential "synchronicity" of phenomena; confronting the repressed "shadow self"; recognizing social roles and "personas"; measurable biofeedback responses to "words" and "ideas" carrying emotional turbulence; and the idea of an *Individuated Self*, the central "archetype" of an individual, underlying all other artificial or added qualities.

## FRANZ BARDON 1909 – 1958

Franz Bardon was the most accomplished *"white wizard"* of the 20th Century. His work effectively serves as an alternative to the training and methods of *"black magicians"* gathering in *"dark brotherhoods"* that are more popularly propagated as New Age *"magick"* —such as books and traditions of Aleister Crowley, &tc. Bardon's philosophy is probably more comparable to the work of 19th Century magician, Eliphas Levi.

Bardon's proficiency in the "occult" was of such repute that his services were sought personally by Adolf Hitler during the second *World War*. When he refused, Bardon was imprisoned in a concentration camp for nearly four years. Following his fortunate escape, he was arrested in Czechoslovakia for failing to pay taxes on alcohol used for his tinctures. He soon after died of "questionable circumstances" in a prison hospital.

During his brief life, Bardon communicated the most superior "spiritual-magical training program" as *"Initiation Into Hermetics: The Path of the True Adept"* (1956) — a course of personal development rivaling *any* graded curriculum from "magical lodges" and "secret societies" of the *New Age*. It also proved valuable to the present author during experimental research for developing practical applications of our philosophy.

Key concepts include: that each level or *plane* of existence forms a systematic *"matrix"* for the world below it; an application of *electro-magnetism* to metaphysics; incorporation of *"Akasha"* (the *Akashic* principle) into hermetic magic; circulation of energetic systems; and transference of *"consciousness"* remote from the body.

---

## GENERAL SYSTEMS THEORY "SYSTEMATOLOGY"

General Systems Theory, academically known as *"systematology"* (or *"systemology"*), is a "study of systematization — *systems* and their formation" and how they dynamically relate to, or interact with (affect and/or are affected by) other *"systems."* Unlike more "specialized" sciences, *systems theory* is not as concerned with what "type" of *system* is being treated — emphasizing *holism* over *reductionism.*

The idea for "systemology" unarguably began with the systematization of Mesopotamian civilization and cuneiform-language written communications. However, as a

*"Throughout human history, as our species has faced the frightening, terrorizing, fact that we do not know who we are, or where we are going in this ocean of chaos, it has been the authorities— the political, the religious, the educational authorities—who attempted to comfort us by giving us order, rules, regulations; informing— forming in our minds—their view of reality. To think for yourself you must question authority and learn how to put yourself in a state of vulnerable open-mindedness to inform yourself."*
—Timothy Leary

*"To operate your brain, you must learn how to use your eyes.*
*'Oh say, can you see?' Oh say, can't you see what is being done to your eyes?*
*Who controls your eyeballs, controls your mind, imprints your brain.*
*Oh say, can't you see?—that the messages that hit your eyeballs in modern television are creating realities—imprinting messages from the sponsors who are not interested in your learning how to design your own realities."*
—Timothy Leary

distinct field of study—separate from *"natural philosophy"*—it is a relatively new science, emerging publicly in the post-war era of the 1940's and 1950's.

Academically, origins for the field of "General Systems Theory" (GST) are attributed to Austrian biologist Ludwig von Bertalanffy, who proposed that classical laws (such as thermodynamics) might apply to closed systems, but not necessarily to "open systems" (such as living things). The field quickly expanded with the contributions of others, such as "Cybernetics" (Norbert Wiener, 1948) and "Chaos Theory" (1980's).

Key concepts include: communication (feedback loops and exchanges of information between systems); fractal geometry (repetitive patterns); complex and dynamic systems; sealed or closed systems; "open systems"; and "new systems philosophy" (quantum consciousness and the unified akashic field).

---

## TIMOTHY LEARY 1920 – 1996

Dr. Timothy Leary served as a psychology professor at Harvard in the early 1960's during some of the first psychedelic research in America. But, by 1962, Leary and other staff involved in the experiments—notably Ralph Metzner and Richard Alpert (later known as *Ram Dass*)—were dismissed from Harvard altogether.

After a rejection from academia, Leary and his associates continued their work independently, sparking the *"1960's New Consciousness"* movement. It's philosophy suggested "dropping out" of participation with the industrialized Western world and returning to "Nature." This also coincided with, and even fed, the early start of the "*New Age*" movement that is still quite prevalent today.

An intermix of "Eastern thought" is visibly present in the original *New Consciousness* movement. Leary was also undoubtedly inspired by Carl Jung, whom he pays tribute to in his debut book: "*The Psychedelic Experience*" (1964), a manual incorporating meditations based on the "*Tibetan Book of the Dead.*" Its briefer sequel, "*Psychedelic Prayers*" (1966) is primarily a version of the "*Tao Te Ching.*"

Most of the *exoteric* emphasis on Leary's legacy surrounds the use of "psychedelics" directly. However, the "psychedelics" were really a "tool" used by an intelligent and creative psychologist in order to explore, experience, and communicate, about aspects of the Mind that conventional science failed to understand.

It's important to note that effectively applying information earned from these experiments is not actually dependent on using psychedelics personally. But the theme was certainly a common one among Leary's circle of friends—Allen Ginsberg, Robert Anton Wilson, and Ken Kesey, just to name a few.

Key concepts include: "circuits" of learning; imprinting on the Mind; cellular "genetic memory" (the 'electric chain of remembrance'); *Self-determined* control and operation of the Mind; transhumanism; and systematically restructuring ("changing") the Mind as a therapeutic process.

"There is no reality until that reality is perceived. Our perceptions of reality will, consequently, appear somewhat contradictory, dualistic and paradoxical.
However, the instantaneous experience of the reality of an immediate occurrence will not appear paradoxical at all.
Reality only seems paradoxical when we construct a history of our perception."
—Fred Alan Wolf

"Right now, you have some sense of being present in your body looking out at the world. But according to physics, this is an illusion of perception. There is no place inside your body where 'you' actually exist. You don't have a particular volume of space or spot that is 'you'. It is an illusion to think that everything outside that volume of space is 'not-you', what you commonly say is 'outside of you'. The best description we can give for this sense of presence is that 'you are everywhere. The main reason that you have more awareness of being in a body is simply because the sensory apparatus of the body commands a great deal of your attention and that much of your attention is linked to your physical senses. We have the illusion that our human bodies are solid, but they are over 99.99% empty space."
—Fred Alan Wolf

## DEEPAK CHOPRA

Deepak Chopra (born in 1946) is one of the few significant influencers of *Systemology* (albeit unknowingly) that is still living at the time of this writing. And again, we see "Westernization of Eastern philosophy and mysticism" as the primary staple defining a modern contribution. Chopra's philosophy is also significantly influenced by Jiddu Krishnamurti (a noted *Theosopher*).

Since 1995, Chopra's *"The Way of the Wizard: Twenty Spiritual Lessons for Creating the Life You Want"* has remained at the very top of the present author's "suggested reading" list for students seeking supplemental outside resources. This book also inspired several popular PBS-TV specials around the time of its release.

Key concepts include: freeing *Self* from the pageantry of playing "roles" in life; *intentions* directing *Awareness*; clearing emotional fragmentation and counter-intuitive thoughts; clearing rigid associations with labels; existence of a "true" or "undefiled" *Self* beneath the artificial attachments and additions.

## 21ST CENTURY METAPHYSICS

The present author began compiling material in the 1990's for what is now *Systemology*—but it would take nearly a quarter-of-a-century before it could be adequately communicated. Of course, in the meantime, the world kept turning, as they say. Others were—more actively—also ushering in a new millennium *paradigm*.

In 2004, independent theaters across the country screened a documentary-like film titled: *"What the Bleep Do We Know?"* The film follows a basic modern-day storyline to graphically illustrate some of its points, but it really surrounds various clips taken from interviews with spiritual-minded scientists and physicists.

Fred Alan Wolf is one of the quantum physics professors interviewed, but an important one. His book *"Taking the Quantum Leap: The New Physics for Non-Scientists"* (1982) was the present author's literary introduction to the subject. It is "recommended reading" for those interested in understanding such better. Dean Radin and William Tiller are just a few of the other relevant individuals interviewed for the film.

The internet age has brought with it a whole assortment of *cyber-gurus* and dot-com enlightenment schools. Even the *Mardukite Org* and *Systemology Society* started much in this way on the internet in 2008—and there are more than a few folk (often with greater resources for publicity) that piggy-backed on the example we set. Nonetheless we have pushed through and stood the test of time to be here with you now *fifteen years later!*

There is much talk of using all this information to *"change the world!"*—yes, we hear that battle call too often to *"change the world!"* We tend to forget that it is us that is *creating* this world; and if anything, it is *us* that requires the "change" and "defragmentation," the "clearing" and "alchemical transformation" of *Self* back into *gold*.

The data on which our *Systemology* is based has "been there all along"—"hidden in plain sight." Much like *General Systems Theory* and *Game Theory* before us, we have simply decided to rearrange the importance and reexamine the significances of available data that might just be useful for broad wide-scope high-view applications. And there is certainly no higher application than the *Great Work*, the journey on a '*Pathway*' that leads to our *Spiritual Ascension*—our return to the *Source*.

"*Humans believe that they are physical machines*
*that learned to think.*
*Actually they are thoughts*
*that learned to create a physical machine.*"
—Deepak Chopra

"*You have changed the past by*
*no longer letting it influence*
*actions in the present.*"
—Deepak Chopra

"*Live the highest ideal now...*
*You do not have to achieve these states*
*in order to live them now.*
*Living them now is how you achieve them.*"
—Deepak Chopra

"*After you have reached the higher stages of the*
*journey, you will be glad to discard all of the*
*clothing with which you have covered the Spirit,*
*and finally Self will stand forth on higher planes*
*naked and be not ashamed.*"
—William Walker Atkinson

*"All that we see or seem
is but a dream within a dream."*
—Edgar Allan Poe

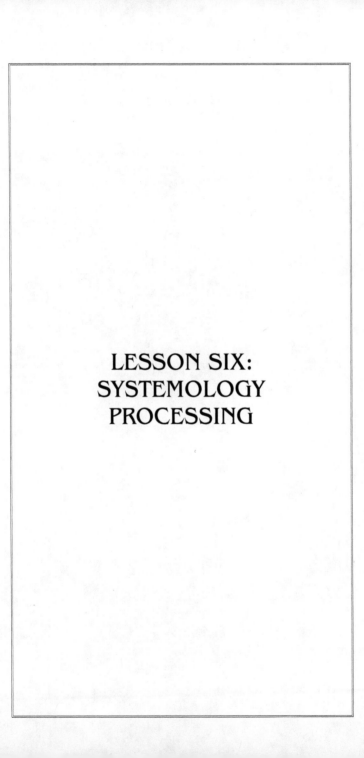

# LESSON SIX:
# SYSTEMOLOGY
# PROCESSING

## PRACTICES OF SPIRITUAL AWAKENING

The philosophy of *Systemology* may be applied to many fields and areas of everyday life. When *applied* within our tradition, practices and exercises toward "spiritual awakening" — or the *"Pathway to Ascension"* — are referred to as *"systematic processing."* Some of the inspiration behind this is derived from sources highlighted in *"Lesson 5."* But really this is the product of many years and innumerable sources.

*"Systematic Processing"* is only *one* part of our philosophy. It is itself an entire "methodology" for applying *Systemology* as a personal practice of techniques and exercises. We call it *"systematic processing"* because it is a precise practice (or "ritual") that *knowingly* mirrors or duplicates the "systematic processes" of the *Mind*. It is too broad a topic to cover fully in a *Basic Course* lesson (booklet), but we can introduce its practice.

One of the basic goals of *"processing"* is to treat *knowingly* (at an analytical level of *awareness*) what is happening "automatically" or "compulsively" — or otherwise *unknowingly*. However, *processing* may be better understood as techniques aimed toward "spiritual development" and "ability enhancement." Of course, these *processes* or techniques are *applications* based on our *philosophy*. Hence: "applied philosophy."

Effectiveness of *systematic processing* is entirely based on our principle axiom that: The "I" or *Self* is an *Alpha-Spirit* operating from a *Spiritual ("Alpha") Existence* and employing a *Mind-System* to perceive the sensory experience of a *genetic-vehicle* (or *'body'*) that interacts in a *Physical ("Beta") Existence*. This axiom *is* "The" *Fundamental of Systemology*—as already explored throughout the previous lessons (booklets).

*Systematic Processing* does resemble some practices from our predecessors in history; perhaps because it reaches for some of the same goals and ideals. It is, however, *not* synonymous with (or equivalent to) these other methods —meditation, prayer, mental healing, therapy, psychoanalysis, *&tc.*—at least not as they are commonly understood by the general (*exoteric*) public. Therefore, we tend to avoid using such terminology.

The most common application of *Systemology* is "*defragmentation*"—but this is only *one* use of "*processing.*" *Defragmentation* is best understood with the model of "*knowns* versus *not-knowns*" given in the introduction. While many assume this means only a *knowledge* of "facts," the same example could illustrate other uses of *processing*, such as a gradual increase of "dormant" (or forgotten) "*spiritual ability*" into "*actual ability*" *&tc.*

---

## BASIC METHODS OF PROCESSING

An "applied philosophy" is really only as effective as an individual *understands* that philosophy and is able to *apply* it in practice. If a *Seeker* is unable to *understand* the philosophy enough to *apply* it, the logical solution is to find an individual that is professionally trained to *understand* it, until the *Seeker* has a *reality* on it themselves. For this, we developed the idea of a *"Pilot"* that could assist in guiding a *Seeker* on the *Pathway*.

This *"Pilot"* concept led to three basic "methods" of *systematic processing*, originally distinguished as *Piloted, Co-Piloted* and *Flying-Solo*. Over the course of many years of additional work, the meaning implied by these terms has evolved slightly during development. But, to briefly describe, they are:

*Piloted Processing*—an untrained *Seeker* is processed by a professionally trained *"Pilot."*

*Co-Piloted Processing*—two *Seekers* in training take turns processing one another; or a *Seeker* is processed by a trusted friend, reading from a book.

*Flying-Solo*—a *Seeker* in training processes themselves.

When first established in the 2010's, all *systematic processing* was intended to be *"Piloted"*—administered by a book-trained or Academy-trained *"Pilot"*—or *"Co-Piloted"* with a friend, or fellow *Seeker*, as *"Pilots-in-Training."* And this is exactly how *"Route-1"* (the first experimental method of *processing*) is presented in *"Tablets of Destiny Revelation"* (2019). More recently, *"Co-Piloted"* seems to apply to anything not done *"Solo."*

Soon after our debut publication, we realized the obvious limitations of a *Piloted* approach. We decided that when developing additional routes, they would need to apply to both traditional *"Piloting"* and solitary practitioners (who were *"Flying Solo"*). That being said, there *are* some *processes* (such as *"Route-1"*) that greatly benefit from having another individual present; yet others are just as productive when "run" *Solo*.

A *Seeker* working alone has only the option of being their own *Solo Pilot*. As such, they are responsible for all the "training" as a *Pilot*, and also managing the "*processing session*" for themselves as a *Seeker*. A certain level of *Self-Determinism* and *Actualized Awareness* must already be in place in order to successfully "*Fly Solo*," since no one else is *present* to help maintain a *Seeker's* "presence" (*attention* and *Awareness*) *in session*.

The phrase "*Self-Processing*" is sometimes used in place of "*Solo*"—but this is not necessarily the most accurate term. Whether a *Seeker* is practicing alone or not, all *systematic processing* techniques in *Systemology* are "*Self-processed*"—which is to say, they must be *processed* by *Self*. And if it seems to you that *Systemology is* sometimes talking about software programming, or operating a computer, you're not alone; but it works.

A *Pilot* may assist in directing attention—or "redirecting" attention, if it strays—but the "command line" (verbal instruction) of a *process* is not a "magic spell" that "does something" by itself. To be effective, the *Seeker* must actually *apply* the "command line" of the technique as *Self* to *Self*. This is what we mean by "running a process"—because essentially, a *Seeker* is *processing* the "command" and resulting *data* as *Self*.

---

## SYSTEMATIC PROCESSING SESSIONS

New "*realizations*" and increased *Actualized Awareness* improve a *Seeker's* handling of everyday life. This may be enhanced through educational training, but a more effective means of reaching these higher ideal states is by combining book-learning with actual practice of the *Systemology* techniques and experiencing the exercises we refer to as *"systematic processing."* A few of these have already appeared in the *Basic Course.*

To be strictly technical, the practice of *systematic processing* is conducted in a formal *"session."* We prefer the term "session" because of the common mystical and/or religious connotations associated with the word *"ritual"*—or even *"meditation."* Our use of the phrase *"processing session"* is most accurate since it implies a specific duration of uninterrupted time set aside to focus on a particular procedure—or *process.*

Rather than conceiving the idea of a "session" as being similar to some "hourly counseling" in some other tradition, a "session" is a period of time for "running a process" toward an intended result or "end-point." A single "session" could last twenty minutes or two hours or more (taking breaks as needed, of course).

Proper procedures for conducting a "session" are followed whether a *Seeker* is *Solo* or *Co-Piloted.* This allows a *Seeker* to treat *systematic processing* as its own unique activity. For example: *"bathing"* is its own unique activity; but it also carries with it a whole regimen of "steps" that are followed in sequence—and which eventually become *"routine"* to the *process.* A *Seeker* practices to achieve that same level of familiarity.

That being said: even the fundamental practice of "starting a session," providing *"presence"* (*attention* and *Awareness*) to be "in session," then formally "ending the session," is a *systematic procedure* or *routine* in itself. To ensure effective processing, we add a step to getting *"in* session." A *Seeker* "scans through daily life" for upsets or problems that are "holding" parts of their *attention* or *Awareness*, even if only on the periphery.

When *things* are not handled analytically—or "above the surface"—they remain in suspension. A *thing* does not *disappear* by our withdrawing from it; it waits around, albeit out of view, to be seen *"As-It-Is."* Many aspects of living out the *Human Condition* have a tendency to keep our *attention* suspended. Even when we go about performing other actions, the totality of our available *Awareness* is not generally present.

For example: we might be preparing a meal for our family, but at the same time we are thinking about problems at work, difficulties in managing the bills, and the driver that almost collided with us on the way home. It is also at these moments, when not living deliberately, that accidents can occur—because, the individual "isn't there" or their "mind isn't on it." Either way, they have withdrawn part of their *"presence."*

Applying *"presence"* to a session is one of the most critical fundamentals of *processing*. Without it, a *Seeker* is not actually *"present"* to be *processed*. We aren't interested in *processing* a "body" or "computer-mind"; we are communicating directly with *Self*, the *Alpha-Spirit* that *utilizes* a *"Mind"* and *"Body"*—and *Self* maintains a "spiritual identity" that is independent of a *"Mind"* and *"Body."* It is *Self* that we want *present* in session.

*Systematic Processing* operates by taking *knowing control* of an individual's available *Actualized Awareness* and then increasing it—both the level of *control* and the available *Awareness*. This is similar to taking a small amount of *cer-*

*tainty* or *ability* and "building up" from there. For a *processing session* to be effective, a *Seeker* cannot be withdrawing their *"presence"*—distracted by the upsets of daily life. This must be handled first.

It is no great mystery that having part of our *attention* on all of the aspects of living (that we hold at length on the peripheral boundary of our conscious *Awareness*) might inhibit our focused concentration or the achievement (or experience) of other "states." But, we can't simply ignore this fact, telling our *Seekers* and readers that we hope they feel better one day and come back to us when they do. We handle it *in-session*.

## HANDLING IN-SESSION PRESENCE

Establishing *"in-session presence"* is not simply a preliminary step; it is a *systematic procedure* in itself. In fact, it is actually the *single* element underlying all other practices —of meditation, prayer, ritualism, *&tc.*—that produces any real *effects*. Of course, in other practices, these *effects* are misappropriated to some other *cause*.

The "location"—the environment or "setting"—is one of the first things to consider for establishing a session. This may be in an uninterrupted space outdoors, or in a quiet room. On occasion, the instruction or "command line" of a *process* may be directed toward a particular environmental focus—such as an "item" *in the room*, or a "person" engaged with others in a *public place*. These are indicators of the intended *setting*.

With the *setting* identified, the next step is making certain a *Seeker* is comfortable and relaxed within that environment. It is unproductive to attempt *processing* in an environment that is actively a source of "turbulence" — or worse, the very type of turbulence that is going to be *processed*. There is no reason for some cult-like disconnection; but a *Seeker* should have a *"retreat"* available to them while regaining balance.

Once a "safe location" is available, our next concern is the *Seeker's* comfort and familiarity with the material and intended practice. In traditional *Piloting*, this would include not only a confidence in the philosophy, but also the individual (*"Pilot"*) administering it. When applied to a *Seeker* that is not also in-training, the total control of *processing* must be handled by a *Pilot* until the *Seeker* assumes responsibility for that control.

All of these factors boil down to what we consider the first elements of *"in-session presence"* that a *Seeker* provides to their exercises (*processing*) — the *attention* that they have available and are willing to provide to the session *in* present-time. And when we say *"presence,"* we mean, quite literally, the *Awareness* of actually *"being present"* in the session. This *"presence" is* the unifying factor of all effective spiritual practices.

Other factors being considered, the actual "orientation" of a *Seeker's presence* — present-time *Awareness* — *with* the present space-time *setting* or environment, is the primary "opening procedure" for a *processing session*. An integral part of this procedure is determining if there are any aspects from daily life that are *inhibiting presence*. Sometimes the handling of these aspects alone is enough to qualify a complete session.

A *Seeker* handles the upsets or energetic-turbulence that is already present—or in stimulation—on the *ZU-Line* before attempting to *resurface* or reveal additional layers of *fragmentation*. Energy is always relayed as a "communication" (even internally for our "*perceptions*"); therefore, most upsets can be categorized as a type of "communication-breakdown" or "break in reality"—an unexpected interruption in the energy *flow*.

Different *processing* techniques target specific aspects of life. But, as a general rule—and in regards to *presence*—the basic idea is to *confront* ("come around to face") or *reach* rather than *avoid* or *withdraw*. Those *things* that we have 'blanked out'—or choose not to look at—are still "energetic masses" surrounding our "field" of *Awareness*. All a *Seeker* really needs to do is "*acknowledge*" they exist so *attention* can shift off them.

To approach this *systematically*, a *Seeker* "spots" whatever incident or aspect is bothering them, then *looks* it over carefully and *analytically*. This may be done as thoroughly as is needed—noticing various things about it, but essentially *acknowledging* or *confronting* a thing "*As-It-Is.*" This allows a *Seeker* to put some distance between themselves and the energetic-mass ("*problem*") rather than treating it as though it were "present."

We return to our previous example of worrying about things while preparing a meal, even if *unknowingly*, which leads to an accident. What is taking place is that the individual is treating those other things as though they are *present* in the environment—giving them "attention- units" as though they are an imminent threat. Of course, by agreement, they are *real things, real incidents*, but *are* they actually "*present*" in session?

Once a troublesome aspect or incident is analyzed (as above), there should be some feeling of relief and an ability to refocus *attention* and *presence in-session*. A "problem" should *seem* "further" away rather than overwhelmingly close. If this is not the case, there may be another upset or distraction. If the troublesome aspect *seems* closer, then alternate *looking* at something in the incident and something in your environment.

This brings us to the other part of establishing *presence in-session*, which is "orientation with the present space-time" (or environment). Essentially, this is a part of what a *Seeker* is employing when *attention* is alternated by *looking* at something in the incident or situation and then something in the physical surroundings. We employ a similar technique as an "opening procedure" for all formal sessions as well.

In the case of handling "problems" or upsets, the alternation allows a *Seeker* to "unfix" *unknowing attention* on something that is being "fixedly" treated as "present" (*in-the-now*) when it is really *somewhere* else or a *past* situation no longer happening presently. These aspects that *attention* is *unknowingly* "fixed" to are what reduces available *Actualized Awareness* that is applied to any "present-time" activity—including *processing*.

Once *Awareness* is able to be focused or concentrated, this additional step—"orientation with present space-time" is essentially exactly what it sounds like: "orienting" *Self* in the present time and space of the session. This means bringing total available *Awareness* to the present-time location (space) of the session and the control of the "body" *in-session*. With regular practice these steps take less time further along the *Pathway*.

The experience of *"Presence In-Session"* will be familiar to continuing students and readers. Versions of sample techniques used as "opening procedures" of a *"Formal Session"* are given in the previous *Basic Course* lessons (booklets) as follows:

*Lesson 1*, Exercise 1 and 2;

*Lesson 2*, Exercise 1 and 7;

*Lesson 3*, Exercise 1, 2 and 7.

## THE FORMAL SESSION

All *systematic processing* is practiced as part of a *"formal"* *systematic processing session*. Of course, there are some exercises—such as those included at the end of each lesson (booklet)—that are effective even when practiced on their own. However, whether *Solo* or *Co-Piloted*, a *Seeker* applying our methods as a total *Pathway* toward *Spiritual Awakening* (or *"Ascension"*), benefits most by practicing techniques *systematically*.

A session is considered *"formal"* because it follows a specific pattern or ceremonial formula of *formal* action and communication. There is a *formal* "beginning" and "ending" of a *systematic session*. Several *processes* may occur within a single session; each one "started" and "stopped" (in turn) as a *formal* act.

Traditional *Piloted Processing* differs from *Flying-Solo*, but the basic formula for a *"Formal Session"* is the same. In *Piloted* (or *Co-Piloted*) processing, the factor of *communication* between a *Pilot* and *Seeker* must be handled in addition to the *session* and *processes*. In *Solo-Processing*, the *communication* of a *process in-session* is all handled "internally"—between *Self* and the *"Mind"* (and *"Body"*) without being directed by a *"Co-Pilot."*

A precise instruction for a *process* is referred to as a *"processing command line"* (or *"PCL"*). This is named for the act of "inputting" a "command line" into a "computing device." In traditional *Piloted* sessions, a *Seeker* receives a communicated "command line" from the *Pilot*, then communicates the "command" to the *Mind* as *Self. Solo* or not, a *Seeker* directs their own *Mind* to *process* the "command"; another *Pilot* only assists this.

A *"processing command line"* (PCL), by itself, is *not* a "magical incantation" that spontaneously produces *actualization.* A misunderstood one, however, *does* have the "power" to slow or stop forward progress. A *Seeker* that is studying/training on their own usually does not have an issue *in-session*, because they can get a clear understanding beforehand. A *Pilot* is responsible for making sure of this for a *Seeker* not in-training.

For example: if you take someone randomly off the street and ask, *"would it be okay if we start this session?"* Well, there is obviously little context there. Even an individual that is interested in getting *processed* might not understand the use of the word "session" at first unless some of the basic *Systemology* philosophy is explained. A *Solo-Pilot* simply accepts total responsibility for attaining this understanding for themselves.

*Solo-Sessions* may be run *"silently"* as direct "mental commands"—but this is not an absolute rule. In either case, a *Seeker* should be focused on handling the commands directly and not *imagine* also *being* "another person" that is giving themselves the commands. Such "add-ons" are unnecessary and counter-productive. However, the same formal session "script" that a *Co-Pilot* communicates, a *Solo-Pilot* reads and *Self-Directs*.

The whole purpose of a *processing session* surrounds the idea of an individual focusing and increasing their *Self-directed* control—or *Self-determinism*. In order to retain this focus, every *one* PCL of a formal session or *process* must have a *Seeker's* total *presence*. This level of focused direction is one benefit of traditional *Piloting*—but in *Solo*, a *Seeker* is instructed to keep a piece of paper over portions of a script/process not yet used.

What follows is a basic script from a *Formal Session* used for training purposes at the Academy. It is a guideline only—based on a transcript of a traditional *Piloted Processing Session*. It is, however, easily adapted to use for *Flying-Solo* once a *Seeker* has practiced the exercises used for achieving *presence in-session*.

### 1. BEGINNING THE SESSION
*"Would it be okay with you if we begin this session now?"*
*"Okay."*
*"Start of session."*

### 2. OPENING PROCEDURES
 **A.** Presence In-Session
*"Is there anything going on that might keep your attention from being present in-session?"*

(if *"no,"* acknowledge and go to *B.*; if *"yes,"* continue below)

*"Okay. Tell me about it."*
*"Alright. How does that problem seem to you now?"*

(if *"further away"* or handled, acknowledge and go to *B.*; if *"closer"* or more turbulent, continue below)

*"Spot something in the incident; Spot something in the room."*

(this alternating command line is repeated as needed)

**B.** Orientation in Present Space-Time

*"Get the sense of you making that body sit in that chair."*

*"Okay. Get a sense of the floor beneath your feet."*

*"Do you have that real good?"*

(if *"no,"* acknowledge and repeat *A.*; if *"yes,"* continue below)

*"Recall a time something seemed real to you."*

*"Tell me something you notice about it."*

*"Look around and spot something in the room."*

*"What do you notice about that?"*

(these last four command lines are repeated in series as needed; acknowledge and continue below)

**C.** Control of Body and Mind In-Session

(two dissimilar objects—here given as *"Item-1"* and *"Item-2"*—are presented and placed within reach; or alternatively, at two distant points in the room, in which a command line for "walking" between them would be inserted)

*"Pick up Item-1."*

*"Tell me about its weight."*

*"Tell me about its color."*

*"Tell me about its texture."*

*"Put it down."*

*"Pick up Item-2."*

*"Tell me about its weight."*

*"Tell me about its color."*

*"Tell me about its texture."*

*"Put it down."*

(this series of command-lines may be repeated several times; when there is no communication-lag for several full series, and duplicate answers are reoccurring, acknowledge and continue below)

*"Choose an object. Decide when you are going to reach for it. Then make that body pick it up."*

*"Now decide when you are going to put it down. Then make that body put it back where it was."*

(repeat as needed; when there is no communication-lag for a full series of command lines, acknowledge and continue below)

*"Close your eyes. Put all of your attention on the upper two back corners of the room and just get real interested in them for a while."*

(if there are no visible signs of "strain" after two minutes, acknowledge and continue below)

*D.* Establishing the Session

*"Do you have any goals for this session, or anything in particular you want to address?"*

(acknowledge, then start a process)

### 3. STARTING A PROCESS

*"I would like to start a process; would that be okay?"*

*"Alright. The command lines are ---. Does this make sense?"*

(if *"no,"* clear up any misunderstood words; if *"yes,"* start the process)

### 4. CHANGING A PROCESS

(only the wording in a command line may be changed to make it more workable for a *Seeker*; to change processes altogether, the present process must reach an end-point)

Example: a Seeker expresses inability to "imagine" or visualize imagery.

*"Okay. Well, just 'get a sense' of..."* or *"Just 'get the idea' of..."*

Example: a Seeker expresses discomfort (or withdrawal from) recalling a particular incident.

*"That's fine. What part of that incident 'could' you confront?"*


**5.** <u>STOPPING A PROCESS</u>

(when an end-point has been reached on a repetitive-style process)

*"We'll just run this process a couple more times if that's okay with you?"*

(general process is run two more times)

*"Okay. Is there anything you would like to tell me before we end this process?"*

(**or**, if an end-point "realization" is communicated from a process)

*"Alright. Very good."*

(the formal end of a particular process requires a command-line)

*"End of process."*


**6.** <u>ENDING THE SESSION</u>

(once a process, or series of processes, is completed)

*"Is there anything you would like to tell me before we end this session?"*

(if *"yes,"* acknowledge and handle it with communication before ending the session; if *"no,"* continue below)

*"Would it be okay if we ended this session now?"*

*"Okay. End of session."*

## YOUR FUTURE AND SYSTEMOLOGY

The *Systemology Society* has established several "levels" of *processing* that are available for study and practice. There are literally *hundreds* of techniques and *systematic processes* that a *Seeker* may apply to their own journey on the *Pathway to Spiritual Ascension*. This *Basic Course* allows that additional study to be more effective; because all the *processes* in the world do not replace a *Seeker's* ability to provide *presence in-session*.

*Seekers* that have completed the *Basic Course* series may be wondering what to study and practice next. The new "*Professional Course*" series provides training and processing for the entire *Pathway* leading up to the upper-most levels of our *Systemology* work. It consists of approximately *30* progressive lessons (booklets)—similar in style to the presentation of the *six* in this *Basic Course* series on "*Fundamentals of Systemology.*"

## PRACTICE EXERCISES

1. Practice speaking the command lines of a *Formal Session*. This will familiarize yourself with the flow of a *Formal Session*, even for *Solo-Processing*. Do not actually "perform" the steps at this point; simply read the lines and speak them out loud. Start with only Step-1 *and* Step-6 ("Beginning" and "Ending" a *Formal Session*). Then add each part of Step-2 ("Opening Procedures") one at a time. [The parts of Step-2 are distinguished by letters A, B, C and D.] Finally, you may include the examples for the remaining steps regarding handling of a particular process, thereby rehearsing the entire script.

2. Practice speaking the command lines for Step-1 *and* Step-6 ("Beginning" and "Ending" a *Formal Session*); but this time with actual intention. Get the idea or sense of actually putting yourself into, and coming out of, a *Formal Session*. Do not "perform" any additional steps at this point. Simply get the idea of setting aside a particular period of time in space to conduct a *Formal Session*.

3. Refer to the previous exercise; but this time *silently* use intention to "Begin" and "End" a *Session*.

4. In this exercise: "Begin" (Step-1) a Formal Session for yourself; then practice performing each part (letter) of the "Opening Procedure" (Step-2)

as a Solo exercise in its own session until you are satisfied. Be sure to include the "End" (Step-6) for each practice session. Refer to the material in this *Basic Course* lesson (booklet) for assistance on achieving "*Presence In-Session.*"

5. Practice communicating a *Formal Session* with actual intention, to a friend or family member. Use Step-1; Step-2, A, B, C; Step-5 (for when an *end-realization* has occurred); and Step-6. No other steps or processes need to be run for this exercise. This practice should be considered a *real* exercise; and you can encourage the other person to actually perform the procedure. Even without the inclusion of additional processing, this basic "regimen" or "routine" has the potential to provide beneficial results—and this includes individuals not studying, or otherwise practicing, *Systemology.*

# SYSTEMOLOGY GLOSSARY

**A-for-A (one-to-one)** : an expression meaning that what we say, write, represent, think or symbolize is a direct and perfect reflection or duplication of the actual aspect or thing—that "A" is for, means and is equivalent to "A" and not "a" or "q" or "!"; in the relay of communication, the message or particle is sent and perfectly duplicated in form and meaning when received.

**actualization** : to make actual, not just potential; to bring into full solid Reality; to realize fully in *Awareness* as a "thing."

**affinity** : the apparent and energetic *relationship* between substances or bodies; the degree of *attraction* or repulsion between things based on natural forces; the *similitude* of frequencies or waveforms; the degree of *interconnection* between systems.

**agreement (reality)** : unanimity of opinion of what is "thought" to be known; an accepted arrangement of how things are; things we consider as "real" or as an "is" of "reality"; a consensus of what is real as made by standard-issue (common) participants; what an individual contributes to or accepts as "real"; in *Systemology*, a synonym for "*reality.*"

**alpha** : the first, primary, basic, superior or beginning of some form; in *Systemology*, referring to the state of existence operating on spiritual archetypes and postulates, will and intention "exterior" to the low-level condensation and solidity of energy and matter as the 'physical universe'.

**alpha-spirit** : a "spiritual" *Life*-form; the "true" *Self* or I-AM; the *individual*; the spiritual (*alpha*) *Self* that is animating the (*beta*) physical body or "*genetic vehicle*" using a continuous *Lifeline* of spiritual ("*ZU*") energy; an individual spiritual (*alpha*) entity possessing no physical mass or measurable waveform (motion) in the Physical Universe as itself, so it animates the (*beta*) physical body or "*genetic vehicle*" as a catalyst to experience *Self*-determined causality in effect within the *Physical Universe*; a singular unit or point of *Spiritual Awareness* that is *Aware* that it is *Aware.*

**alpha thought** : the highest spiritual *Self-determination* over creation and existence exercised by an Alpha-Spirit; the Alpha range of pure *Creative Ability* based on direct postulates and considerations of *Beingness*; spiritual qualities comparable to "thought" but originating in Alpha-existence (at "6.0") independently superior to a *beta-anchored* Mind-System, although an Alpha-Spirit may use Will ("5.0") to carry the intentions of a postulate or consideration ("6.0") to the Master Control Center ("4.0").

**apparent** : visibly exposed to sight; evident rather than actual, as presumed by Observation; readily perceived, especially by the senses.

**archetype** : a "first form" or ideal conceptual model of some aspect; the ultimate prototype of a form on which all other conceptions are based.

**ascension** : actualized *Awareness* elevated to the point of true "spiritual existence" exterior to *beta existence*. An "Ascended Master" is one who has returned to an incarnation on Earth as an inherently *Enlightened One*, demonstrable in their actions—they have the ability to *Self-direct* the "Spirit" as *Self* and maintain consciousness beyond this existence as a personal identity continuum with the same *Self-directed* control and communication of Will-Intention that is exercised, actualized and developed deliberately during one's present incarnation.

**assessment** : an analysis or synthesis of collected information, usually about a person or group, in relation to an *assessment scale.*

**associative knowledge** : significance or meaning of a facet or aspect assigned to (or considered to have) a direct relationship with another facet; to connect or relate ideas or facets of existence with one another; a reactive-response image, emotion or conception that is suggested by (or directly accompanies) something other than itself; in traditional systems logic, an equivalency of significance or meaning between facets or sets that are grouped together, such as in $(a + b) + c = a + (b + c)$; in Systemology, erroneous associative knowledge is assignment of the same value to all facets or parts considered as related (even when they are not actually so), such as in $a = a,\ b = a,\ c = a$ and so forth without distinction.

**attention** : active use of *Awareness* toward a specific aspect or thing; the act of "attending" with the presence of *Self*; a direction of focus or concentration of *Awareness* along a particular channel or conduit or toward a particular terminal node or communication termination point; the Self-directed concentration of personal energy as a combination of observation, thought-waves and consideration; focused application of *Self-Directed Awareness*.

**awareness** : the highest sense of-and-as Self in knowing and being as I-AM (the *Alpha-Spirit*); the extent of beingness directed as a POV experienced by Self as knowingness.

**axiom** : a fundamental truism of a knowledge system, esp. *logic*; all *maxims* are also *axioms*; knowledge statements that require no proof because their truth is self-evident; an established law or systematic principle used as a *premise* on which to base greater conclusions of truth.

**beta (awareness)** : all consciousness activity ("*Awareness*") in the "Physical Universe" (KI) or else *beta-existence*; *Awareness* within the range of the *genetic-body*, including material thoughts, emotional responses and physical motors; personal *Awareness* of physical energy and physical matter moving through physical space and experienced as "time"; the *Awareness* held by *Self* that is restricted to a physical organic *Lifeform* or "*genetic vehicle*" in which it experiences causality in the *Physical Universe*.

**beta (existence)** : all manifestation in the "Physical Universe" (KI); the "Physical" state of existence consisting of vibrations of physical energy and physical matter moving through physical space and experienced as "time"; the conditions of *Awareness* for the *Alpha-spirit* (*Self*) as a physical organic *Lifeform* or "*genetic vehicle*" in which it experiences causality in the *Physical Universe*.

**beta-defragmentation** : toward a state of *Self-Honesty* in regards to handling experience of the "Physical Universe" (*beta-existence*); an applied spiritual philosophy (or technology) of Self-Actualization originally described in the text "*Crystal Clear*" (*Liber-2B*), building upon theories from "*Systemology: The Original Thesis*."

**catalyst** : something that causes action between two systems or aspects, but which itself is unaffected as a variable of this energy communication; a medium or intermediary channel.

**chakra** : an archaic Sanskrit term for "wheel" or "spinning circle" used in *Eastern* wisdom traditions, spiritual systems and mysticism; a concept retained in Systemology to indicate etheric concentrations of energy into wheel-mechanisms that process *ZU* energy at specific frequencies along the *ZU-line*, of which the *Human Condition* is reportedly attached *seven* at various degrees as connected to the Gate symbolism.

**channel** : a specific stream, course, current, direction or route; to form or cut a groove or ridge or otherwise guide along a specific course; a direct path; an artificial aqueduct created to connect two water bodies or water or make travel possible.

**charge** : to fill or furnish with a quality; to supply with energy; to lay a command upon; in *Systemology*—to imbue with intention; to overspread with emotion; application of *Self-directed (WILL)* "intention" toward an emotional manifestation in beta-existence; personal energy stores and significances entwined as fragmentation in mental images, reactive-response encoding and intellectual (and/or) programmed beliefs; in traditional mysticism, to intentionally fix an energetic resonance to meet some degree, or to bring a specific concentration of energy that is transferred to a focal point, such as an object or space.

**circuit** : a circular path or loop; a closed-path within a system that allows a flow; a pattern or action or wave movement that follows a specific route or potential path only; in *Systemology*, "*communication processing*" pertaining to a specific flow of energy or information along a channel; *see* also "*feedback loop.*"

**communication** : successful transmission of information, data, energy (&tc.) along a message line, with a reception of feedback; an energetic flow of intention to cause an effect (or duplication) at a distance; the personal energy moved or acted upon by will or else 'selective directed attention'; the 'messenger action' used to transmit and receive energy across a medium; also relay of energy, a message or signal—or even locating a personal POV (viewpoint) for the Self—along the *ZU-line*.

**compulsion** : a failure to be responsible for the dynamics of control—starting, stopping or altering—on a particular channel of communication and/or regarding a particular terminal in existence; an energetic flow with the appearance of being 'stuck' on the action it is already doing or by the control of some automatic mechanism.

**concept** : a high-frequency thought-wave representing an "idea" which persists because it is not restricted to a unique space-time; an abstract or tangible "idea" formed in the "Mind" or *imagined* as a means of understanding, usually including associated "Mental Images"; a seemingly timeless collective thought-theme (or subject) that entangles together facets of many events or incidents, not just a single significant one.

**condense (condensation)** : the transition of vapor to liquid; denoting a change in state to a more substantial or solid condition; leading to a more compact or solid form.

**condition** : an apparent or existing state; circumstances, situations and variable dynamics affecting the order and function of a system; a series of interconnected requirements, barriers and allowances that must be met; in "contemporary language," bringing a thing toward a specific, desired or intentional new state (such as in "conditioning"), though to minimize confusion about the word "condition" in our literature, *Systemology* treats "contemporary conditioning" concepts as imprinting, encoding and programming.

**conflict** : the opposition of two forces of similar magnitude along the same channel or competing for the same terminal; the inability to duplicate another POV; a thought, intention or communication that is met with an opposing counter-thought or counter-intention that generates an energetic cluster.

**confront** : to come around in front of; to be in the presence of; to stand in front of, or in the face of; to meet "face-to-face" or "face-up-to"; additionally, in *Systemology*, to fully tolerate or acceptably withstand an encounter with a particular manifestation or encounter.

**consciousness** : the energetic flow of *Awareness*; the Principle System of *Awareness* that is spiritual in nature, which demonstrates potential interaction with all degrees of the Physical Universe; the *Beingness* component of our existence in *Spirit*; the Principle System of *Awareness* as *Spirit* that directs action in the Mind-System.

**consideration** : careful analytical reflection of all aspects; deliberation; determining the significance of a "thing" in relation to similarity or dissimilarity to other "things"; evaluation of facts and importance of certain facts; thorough examination of all aspects related to, or important for, making a decision; the analysis of consequences and estimation of significance when making decisions; in *Systemology*, the postulate or Alpha-Thought that defines the state of beingness for what something "*is.*"

**continuity** : being a continuous whole; a complete whole or "total round of"; the balance of the equation [ "–120" + "120" = "0" *&tc.*]; an apparent unbroken interconnected coherent whole; also, as applied to Universes in *Systemology*, the lowest base consideration of space-time or commonly shared level of energy-matter apparent in an existence, or else the lowest degree of solidity or condensation whereby all mass that exists is identifiable or communicable with all other mass that exists; represented as "0" on the *Standard Model* for the Physical Universe (*beta-existence*), a level of existence that is below Human emotion, comparable to the solidity of "rocks" and "walls" and "inert bodies."

**continuum** : a continuous enduring uninterrupted sequence or condition; observing all gradients on a *spectrum*; measuring quantitative variation with gradual transition on a spectrum without demonstrating discontinuity or separate parts.

**control (general)** : the ability to start, change or start some action or flow of energy; the capacity to originate, change or stop some mode of human behavior by some implication, physical or psychological means to ensure compliance (voluntarily or involuntarily).

**control (systems)** : communication relayed from an operative center or organizational cluster, which incites new activity elsewhere in a system (or along the *ZU-line*)

**defragmentation** : the *reparation* of wholeness; collecting all dispersed parts to reform an original whole; a process of removing "*fragmentation*" in data or knowledge to provide a clear understanding; applying techniques and processes that promote a *holistic* interconnected *alpha* state, favoring observational *Awareness* of continuity in all spiritual and physical systems; in *Systemology*, a "*Seeker*" achieving an actualized state of basic "*Self-Honest Awareness*" is said to be *beta-defragmented*, where as *Alpha-defragmentation* is the rehabilitation of the *creative ability*, managing the *Spiritual Timeline* and the POV of *Self* as Alpha-Spirit (I-AM); see also "*Beta-defragmentation*."

**degree** : a physical or conceptual *unit* (or point) defining the variation present relative to a *scale* above and below it; any stage or extent to which something *is* in relation to other possible positions within a *set* of "*parameters*"; a point within a specific range or spectrum; in *Systemology*, a *Seeker's* potential energy variations or fluctuations in thought, emotional reaction and physical perception are all treated as "*degrees*."

**dramatization / dramatize** : a vivid display or performance as if rehearsed for a "play" (on stage); a *'circuit'* recording *'imprinted'* in the past and, once restimulated by a facet of the environment, the individual "replays" it as through reacting to it in the present (and identifying that reality as present reality ); acts, actions and observable behaviors that demonstrate identification with a particular character type, "phase" or personality program; a motivated sequence-chain, implant series or imprinted cycle of actions—usually irrational or counter-survival—repeated by an individual as it had previously happened to them; a reoccurring or reactively triggered out-flow, communication or action that indicates an individual "occupying" a particular *'Point-of-View'* (*POV*)—typically fixed to a specific (past) identification (identity) that is space-time locatable (meaning a point where significant *Attenergy* —enough to compulsively create and maintain a POV—is "stuck" or "hung up" on the *BackTrack*).

**dynamic (systems)** : a principle or fixed system which demonstrates its *'variations'* in activity (or output) only in constant relation to variables or fluctuation of interrelated systems; a standard principle, function, process or system that exhibits *'variations'* and change simultaneously with all connected systems; each *'Sphere of Existence'* is a dynamic system, systematically affecting (supporting) and affected (supported) by other *'Spheres'* (which are also dynamic systems).

**emotional encoding** : the readable substance/material (data) of *'imprints'*; associations of sensory experience with an *imprint*; perceptions of our environment that receive an *emotional charge*, which form or reinforce facets of an *imprint*; perceptions recorded and stored as an *imprint* within the "emotional range" of energetic manifestation; the formation of an energetic store or charge on a channel that fixes emotional responses as a mechanistic automation, which is carried on in an individual's *Spiritual Timeline* (or personal continuum of existence).

**encompassing** : to form a circle around, surround or envelop around.

**end point** : the moment when the goal of a process has been achieved and to continue on with it will be detrimental to the gains; the finality of a process when the *Seeker* has achieved their optimum state from the current cycle (whether or not they run through it again at a later date with a different level of *Awareness* or knowledge base doesn't change the fact that it has flattened the standing wave

**enforcement** : the act of compelling or putting (effort) into force; to compel or impose obedience by force; to impress strongly with applications of stress to demand agreement or validation; the lowest-level of direct control by physical effort or threat of punishment; a low-level method of control in the absence of true communication.

**evaluate** : to determine, assign or fix a set value, amount or meaning.

**existence** : the *state* or fact of *apparent manifestation*; the resulting combination of the Principles of Manifestation: consciousness, motion and substance; continued *survival*; that which independently exists; the *'Prime Directive'* and sole purpose of all manifestation or Reality; the highest common intended motivation driving any "*Thing*" or *Life*.

**exterior** : outside of; on the outside; in *Systemology*, we mean specifically the POV of *Self* that is *'outside of'* the *Human Condition,* free of the physical and mental trappings of the Physical Universe; a metahuman range of consideration; see also '*Zu-Vision*'.

**external** : a force coming from outside; information received from outside sources; in *Systemology*, the objective *'Physical Universe'* existence, or *beta-existence*, that the Physical Body or *genetic vehicle* is essentially *anchored* to for its considerations of locational space-time as a dimension or POV.

**facets** : an aspect, an apparent phase; one of many faces of something; a cut surface on a gem or crystal; in *Systemology*—a single perception or aspect of a memory or "*Imprint*"; any one of many ways in which a memory is recorded; perceptions associated with a painful emotional (sensation) experience and "*imprinted*" onto a metaphoric lens through which to view future similar experiences; other secondary terminals that are associated with a particular terminal, painful event or experience of loss, and which may exhibit the same encoded significance as the activating event.

**faculties** : abilities of the mind (individual) inherent or developed.

**feedback loop** : a complete and continuous circuit flow of energy or information directed as an output from a source to a target which is altered and return back to the source as an input; in *General Systemology*—the continuous process where outputs of a system are routed back as inputs to complete a circuit or loop, which may be closed or connected to other systems/circuits; in *Systemology*—the continuous process where directed *Life* energy and *Awareness* is sent back to *Self* as experience, understanding and memory to complete an energetic circuit as a loop.

**flow** : movement across (or through) a channel (or conduit); a direction of active energetic motion typically distinguished as either an *in-flow*, *out-flow* or *cross-flow*.

**fragmentation** : breaking into parts and scattering the pieces; the *fractioning* of wholeness or the *fracture* of a holistic interconnected *alpha* state, favoring observational *Awareness* of perceived connectivity between parts; *discontinuity*; separation of a totality into parts; in *Systemology*, a person outside a state of *Self-Honesty* is said to be *fragmented*.

**game** : a strategic situation where a "player's" power of choice is employed or affected; a parameter or condition defined by purposes, freedoms and barriers (rules).

**general systemology ("systematology")** : a methodology of analysis and evaluation regarding the systems—their design and function; organizing systems of interrelated information-processing in order to perform a given function or pattern of functions.

**genetic-vehicle** : a physical *Life*-form; the physical (*beta*) body that is animated/controlled by the (*Alpha*) *Spirit* using a continuous *Lifeline* (ZU); a physical (*beta*) organic receptacle and catalyst for the (*Alpha*) *Self* to operate "causes" and experience "effects" within the *Physical Universe*.

**gradient** : a degree of partitioned ascent or descent along some scale, elevation or incline; "higher" and "lower" values in relation to one another.

**holistic** : the examination of interconnected systems as encompassing something greater than the *sum* of their "parts."

**Human Condition** : a standard default state of Human experience that is generally accepted to be the extent of its potential identity (*beingness*)—currently treated as *Homo Sapiens Sapiens,* but which is scheduled for replacement by *Homo Novus*.

**identification** : the association of *identity* to a thing; a label or fixed data-set associated to what a thing is; association "equals" a thing, the "equals" being key; an equality of all things in a group, for example, an "apple" identified with all other "apples"; the reduction of "I-AM"-*Self* from a *Spiritual Beingness* to an "identity" of some form.

**identity** : the collection of energy and matter—including memory—across a "*Spiritual Timeline*" that we consider as "I" of *Self,* but the "I" is an individual and not an identification with anything other than *Self* as *Alpha-Spirit.*

**imagination** : the ability to create *mental imagery* in one's Personal Universe at will and change or alter it as desired; the ability to create, change and dissolve mental images on command or as an act of will; to create a mental image or have associated imagery displayed (or "conjured") in the mind that may or may not be treated as real (or memory recall) and may or may not accurately duplicate objective reality; to employ *Creative Abilities* of the Spirit that are independent of reality agreements with beta-existence.

**imprint** : to strongly impress, stamp, mark (or outline) onto a softer 'impressible' substance; to mark with pressure onto a surface; in *Systemology,* the term is used to indicate permanent Reality impressions marked by frequencies, energies or interactions experienced during periods of emotional distress, pain, unconsciousness, loss, enforcement, or something antagonistic to physical (personal) survival, all of which are are stored with other reactive response-mechanisms at lower-levels of *Awareness* as opposed to the active memory database and proactive processing center of the Mind; an experiential "memory-set" that may later resurface—be triggered or stimulated artificially—as Reality, of which similar responses will be engaged automatically; holographic-like imagery "stamped" onto consciousness as composed of energetic *facets* tied to the "snap-shot" of an experience.

**imprinting incident** : the first or original event instance communicated and *emotionally encoded* onto an individual's "*Spiritual Timeline*" (recorded memory from all lifetimes), which formed a permanent impression that is later used to mechanistically treat future contact on that channel; the first or original occurrence of some particular *facet* or mental image related to a certain type of *encoded response*, such as pain and discomfort, losses and victimization, and even the acts that we have taken against others along the Spiritual Timeline of our existence that caused them to also be *Imprinted*.

**inhibited** : withheld, held-back, discouraged or repressed from some state.

**"in phase"** : see "*phase alignment.*"

**intention** : the directed application of Will; to intend (have "in Mind") or signify (give "significance" to) for or toward a particular purpose; in *Systemology* (from the *Standard Model*)—the spiritual activity at WILL (5.0) directed by an *Alpha Spirit* (7.0); the application of WILL as "Cause" from a higher order of Alpha Thought and consideration (6.0), which then may continue to relay communications as an "effect" in the universe.

**interior** : inside of; on the inside; in *Systemology,* we mean specifically the POV of *Self* that is fixed to the *'internal' Human Condition,* including the *Reactive Control Center* (RCC) and Mind-System or *Master Control Center* (MCC); within *beta-existence*.

**internal** : a force coming from inside; information received from inside sources; in *Systemology*, the objective *'Physical Universe'* experience of *beta-existence* that is associated with the Physical Body or *genetic vehicle* and its POV regarding sensation and perception; from inside the body; within the body.

**invalidate** : decrease the level or degree or *agreement* as Reality.

**knowledge** : clear personal processing of informed understanding; information (data) that is actualized as effectively workable understanding; a demonstrable understanding on which we may 'set' our *Awareness*—or literally a "know-ledge."

**Master-Control-Center (MCC)** : a perfect computing device to the extent of the information received from "lower levels" of sensory experience/perception; the proactive communication system of the "*Mind*"; a relay point of active *Awareness* along the Identity's *ZU-line*, which is responsible for maintaining basic *Self-Honest Clarity* of *Knowingness* as a *seat of consciousness* between the *Alpha-Spirit* and the secondary "*Reactive Control Center*" of a *Lifeform* in *beta existence*; the Mind-center for an *Alpha-Spirit* to actualize cause in the *beta existence*; the analytical *Self-Determined* Mind-center of an *Alpha-Spirit used* to project *Will* toward the genetic body; the point of contact between *Spiritual Systems* and the *beta existence*; presumably the "*Third Eye*" of a being connected directly to the *I-AM-Self*, which is responsible for *determining* Reality at any time; in *Systemology*, this is plotted at (4.0) on the continuity model of the *ZU-line*.

**mental image** : a subjectively experienced "picture" created and imagined into being by the Alpha-Spirit (or at lower levels, one of its automated mechanisms) that includes all perceptible *facets* of totally immersive scene, which may be forms originated by an individual, or a "facsimile-copy" ("snap-shot") of something seen or encountered; a duplication of wave-forms in one's Personal Universe as a "picture" that mirror an "external" Universe experience, such as an *Imprint*.

**methodology** : a complete system of applications, methods, principles and rules to compose a *'systematic'* paradigm as a "whole"—esp. a field of philosophy or science.

**misappropriated** : put into use incorrectly; to apply ineffectively or as unintended by design or definition.

**objective** : concerning the "external world" and attempts to observe Reality independent of personal "subjective" factors.

**one-to-one** : see "*A-for-A.*"

**optimum** : the most favorable or ideal conditions for the best result; the greatest degree of result under specific conditions.

**organic** : as related to a physically living organism or carbon-based life form; energy-matter condensed into form as a focus or POV of Spiritual Life Energy (*ZU*) as it pertains to beta-existence of *this* Physical Universe (*KI*).

**paradigm** : an all-encompassing *standard* by which to view the world and *communicate* Reality; a standard model of reality-systems used by the Mind to filter, organize and interpret experience of Reality.

**parameters** : a defined range of possible variables within a model, spectrum or continuum; the extent of communicable reach capable within a system or across a distance; the defined or imposed limitations placed on a system or the functions within a system; the extent to which a Life or "thing" can *be*, *do* or *know* along any channel within the confines of a specific system or spectrum of existence.

**patterns (probability patterns)** : observation of cycles and tendencies to predict a causal relationship or determine the actual condition or flow of dynamic energy using a holistic systemology to understand Life, Reality and Existence as opposed to isolating or excluding perceived parts as being mutually separate from other perceived parts.

**perception** : internalized processing of data received by the *senses*; to become *Aware of* via the senses.

**personality (program)** : the total composite picture an individual "identifies" themselves with; the accumulated sum of material and mental mass by which an individual experiences as their timeline; a "beta-personality" is mainly attached to the identity of a particular physical body and the total sum of its own genetic memory in combination with the data stores and pictures maintained by the Alpha Spirit; a "true personality" is the Alpha Spirit as Self completely defragmented of all erroneous limitations and barriers to consideration, belief, manifestation and intention.

**phase (identification)** : in *Systemology,* a pattern of personality or identity that is assumed as the POV from *Self*; personal identification with artificial "personality packages"; an individual assuming or taking characteristics of another individual (often unknowingly as a response-mechanisms); also *"phase alignment."*

**phase alignment** or *"in phase"* : to be in synch or mutually synchronized, in step or aligned properly with something else in order to increase the total strength value; in *Systemology*, alignment or adjustment of *Awareness* with a particular identity, space or time; perfect *defragmentation* would mean being "in phase" as *Self* fully conscious and Aware as an Alpha-Spirit *in* present *space* and *time*, free of synthetic personalities.

**physics** : regarding data obtained by a material science of observable motions, forces and bodies, including their apparent interaction, in the Physical Universe (specific to this *beta-existence*).

**physiology** : a material science of observable biological functions and mechanics of living organisms, including codification and study of identifiable parts and apparent systematic processes (specific to agreed upon makeup of the *genetic vehicle* for this *beta-existence*).

**point-of-view (POV)** : a point to view from; an opinion or attitude as expressed from a specific identity-phase; a specific standpoint or vantage-point; a definitive manner of consideration specific to an individual phase or identity; a place or position affording a specific view or vantage; circumstances and programming of an individual that is conducive to a particular response, consideration or belief-set (paradigm); a position (consideration) or place (location) that provides a specific view or perspective (subjective) on experience (of the objective).

**postulate** : to put forward as truth; to suggest or assume an existence *to be*; to state or affirm the existence of particular conditions; to provide a basis of reasoning and belief; a basic theory accepted as fact; in *Systemology*, "Alpha-Thought"—the top-most decisions or considerations made by the Alpha-Spirit regarding the *"is-ness"* (what things "are") about energy-matter and space-time.

**potentiality** : the total "sum" (collective amount) of "latent" (dormant—present but not apparent) capable or possible realizations; used to describe a state or condition of what has not yet manifested, but which can be influenced and predicted based on observed patterns and, if referring to beta-existence, Cosmic Law.

**presence** : the quality of some thing (energy/matter) being "present" in space-time; personal orientation of *Self* as an *Awareness* (*POV*) located in present space-time (environment) and communicating with extant energy-matter.

**Prime Directive** : a "spiritual" implant program that installs purposes and goals into the personal experience of a Universe, esp. any *Beta-Existence* (whether a 'Games Universe' or a 'Prison Universe'); intellectually treated as the "Universal Imperative" in some schools of moral philosophy; comparable to "Universal Law" or "Cosmic Ordering."

**"process-out"** : to reduce *emotional encoding* of an *imprint* to zero; to dissolve a *wave-form* or *thought-formed* "solid" such as a "*belief*"; to completely run a *process* to its end, thereby *flattening* any previous *waves* of *fragmentation* that are obstructing the *clear channel* of *Self-Awareness*; also referred to as "processing-out"; to discharge all previously held emotionally encoded imprinting or erroneous programming and beliefs that otherwise fix the free flow (wave) to a particular pattern, solid or concrete "*is*" form.

**processing, systematic** : the inner-workings or "through-put" result of systems; in *Systemology*, a methodology of applied spiritual technology used toward personal Self-Actualization; methods of selective directed attention, communicated language and associative imagery that targets an increase in personal control of the human condition.

**reactive control center (RCC)** : the secondary (reactive) communication system of the "*Mind*"; a relay point of *Awareness* along the Identity's *ZU-line*, which is responsible for engaging basic motors, biochemical processes and any *programmed automated responses* of a living *beta* organism; the reactive Mind-Center of a living organism relaying communications of *Awareness* between causal experience of *Physical Systems* and the "*Master Control Center*"; it presumably stores all emotional encoded imprints as fragmentation of "chakra" frequencies of *ZU* (within the range of the "*psychological/emotive systems*" of a being), which it may *react* to as Reality at any time; in *Systemology*, this is plotted at (2.0) on the continuity model of the *ZU-line*.

**reality** : see "*agreement*."

**realization** : the clear perception of an understanding; a consideration or understanding on what is "actual"; to make "real" or give "reality" to so as to grant a property of "beingness" or "being as it is"; the state or instance of coming to an *Awareness*; in *Systemology*, "gnosis" or true knowledge achieved during *systematic processing*; achievement of a new (or "higher") cognition, true knowledge or perception of Self; a consideration of reality or assignment of meaning.

**relative** : an apparent point, state or condition treated as distinct from others.

**responsibility** : the *ability* to *respond*; the extent of mobilizing *power* and *understanding* an individual maintains as *Awareness* to enact *change*; the proactive ability to *Self-direct* and make decisions independent of an outside authority.

**resurface** : to return to (or bring up to) the "surface" of that which has previously been submerged; in *Systemology*—relating specifically to processes where a *Seeker* recalls blocked energy stored covertly as emotional "*imprints*" (by the RCC) so that it may be effectively defragmented from the "*ZU-line*" (by the MCC).

**Seeker** : an individual on the *Pathway to Self-Honesty*; a practitioner of *Mardukite Systemology* or *Systemology Processing* that is working toward *Spiritual Ascension*.

**Self-actualization** : bringing the full potential of the Human spirit into Reality; expressing full capabilities and creativeness of the *Alpha-Spirit*.

**Self-determinism** : the freedom to act, clear of external control or influence; the personal control of Will to direct intention.

**Self-honesty** : the basic or original *alpha* state of *being* and *knowing*; clear and present total *Awareness* of-and-as *Self*, in its most basic and true proactive expression of itself as *Spirit* or *I-AM*—free of artificial attachments, perceptive filters and other emotionally-reactive or mentally-conditioned programming imposed on the human condition by the systematized physical world; the ability to experience existence without judgment.

**sensation** : an external stimulus received by internal sense organs (receptors/sensors); sense impressions.

**slate** : a hard thin flat surface material used for writing on; a chalk-board, which is a large version of the original wood-framed writing slate, named for the rock-type it was made from.

**spectrum** : a broad range or array as a continuous series or sequence; defined parts along a singular continuum; in physics, a gradient arrangement of visible colored bands diffracted in order of their respective wavelengths, such as when passing *White Light* through a *prism*.

**Spheres of Existence (dynamic systems)** : a series of *eight* concentric circles, rings or spheres (each larger than the former) that is overlaid onto the Standard Model of Beta-Existence to demonstrate the dynamic systems of existence extending out from the POV of Self (often as a "body") at the *First Sphere*; these are given in the basic eightfold systems as: *Self, Home/Family, Groups, Humanity, Life on Earth, Physical Universe, Spiritual Universe* and *Infinity-Divinity.*

**spiritual timeline** : a continuous stream of moment-to-moment *Mental Images* (or a record of experiences) that defines the "past" of a spiritual being (or *Alpha-Spirit*) and which includes impressions (*imprints, &tc.*) form all life-incarnations and significant spiritual events the being has encountered; in Systemology, also "*backtrack.*"

**standard issue** : equally dispensed to all without consideration.

**Standard Model, The (systemology)** : in *Systemology*—our existential and cosmological *standard model* or cabbalistic model; a "*monistic continuity model*" demonstrating *total system* interconnectivity "above" and "below" observation of any apparent *parameters*; the original presentation of the *ZU-line*, represented as a singular vertical ($y$-axis) waveform in space across dimensional levels or Universes (*Spheres of Existence*) without charting any specific movement across a dimensional time-graph $x$-axis; The Standard Model of Systemology represents the basic workable synthesis of common denominators in models explored throughout Grade-I and Grade-II material.

**static** : characterized by a fixed or stationary condition; having no apparent change, movement or fluctuation.

**succumb** : to give way, or give in to, a relatively stronger superior force.

**system** : from the Greek, "to set together"; to set or arrange things or data together so as to form an orderly understanding of a "whole"; also a *'method'* or *'methodology'* as an orderly standard of use or application of such data arranged together.

**systematization** : to arrange into systems; to systematize or make systematic.

**terminal (node)** : a point, end or mass on a line; a point or connection for closing an electric circuit, such as a post on a battery terminating at each end of its own systematic function; any end point or 'termination' on a line; a point of connectivity with other points; in systems, any point which may be treated as a contact point of interaction; anything that may be distinguished as an 'is' and is therefore a 'termination point' of a system or along a flow-line which may interact with other related systems it shares a line with; a point of interaction with other points.

**thought-form** : apparent *manifestation* or existential *realization* of *Thought-waves* as "solids" even when only apparent in Reality-agreements of the Observer; the treatment of *Thought-waves* as permanent *imprints* obscuring *Self-Honest Clarity* of *Awareness* when reinforced by emotional experience as actualized "thought-formed solids" ("*beliefs*") in the Mind; energetic patterns that "surround" the individual.

**thought-habit** : reoccurring modes of thought or repeated "self-talk"; essentially "self-hypnosis" resulting in a certain state.

**thought-wave** or **wave-form** : a proactive *Self-directed action* or reactive-response *action* of *consciousness*; the *process* of *thinking* as demonstrated in *wave-form*; the *activity* of *Awareness* within the range of *thought vibrations/frequencies* on the existential *Life-continuum* or *ZU-line*.

**threshold** : a doorway, gate or entrance point; the degree to which something is to produce an effect within a certain state or condition; the point in which a condition changes from one to the next.

**tier** : a series of rows or levels, one stacked immediately before or atop another.

**time** : observation of cycles in action; motion of a particle, energy or wave across space; intervals of action related to other intervals of action as observed in Awareness; a measurable wave-length or frequency in comparison to a static state; the consideration of variations in space.

**timeline** : plotting out history in a linear (line) model to indicate instances (experiences) or demonstrate changes in state (space) as measured over time; a singular conception of continuation of observed time as marked by event-intervals and changes in energy and matter across space.

**turbulence** : a quality or state of distortion or disturbance that creates irregularity of a flow or pattern; the quality or state of aberration on a line (such as ragged edges) or the emotional "turbulent feelings" attached to a particular flow or terminal node; a violent, haphazard or disharmonious commotion (such as in the ebb of gusts and lulls of wind action).

**understanding** : a clear 'A-for-A' duplication of a communication as 'knowledge', which may be comprehended and retained with its significance assigned in relation to other 'knowledge' treated as a 'significant understanding'; the "grade" or "level" that a knowledge base is collected and the manner in which the data is organized and evaluated.

**validation** : reinforcement of agreements or considerations as "real."

**viewpoint** : see *"point-of-view" (POV)*.

**will** *or* **WILL** (5.0) : in *Systemology* (from the *Standard Model*), the Alpha-ability at "5.0" of a Spiritual Being (*Alpha Spirit*) at "7.0" to apply *intention* as "Cause" from consideration or Alpha-Thought at "6.0" that is superior to "beta-thoughts" that only manifest as reactive "effects" below "4.0" and *interior* to the *Human Condition*.

**willingness** : the state of conscious Self-determined ability and interest (directed attention) to *Be, Do* or *Have*; a Self-determined consideration to reach, face up to (*confront*) or manage some "mass" or energy; the extent to which an individual considers themselves able to participate, act or communicate along some line, to put attention or intention on the line, or to produce (create) an effect.

*ZU* : the ancient Sumerian cuneiform sign for the archaic verb —*"to know," "knowingness"* or *"awareness"*; in *Mardukite Zuism and Systemology*, the active energy/matter of the "Spiritual Universe" (AN) experienced as a *Lifeforce* or *consciousness* that imbues living forms extant in the "Physical Universe" (KI); "*Spiritual Life Energy*"; energy demonstrated by the WILL of an actualized *Alpha-Spirit* in the "Spiritual Universe" (AN), which impinges its *Awareness* into the Physical Universe (KI), animating/controlling *Life* for its experience of *beta-existence* along an individual Alpha-Spirit's personal *Identity-continuum*, called a *ZU-line*.

**Zu-Line** : a theoretical construct in *Mardukite Zuism and Systemology* demonstrating *Spiritual Life Energy* (*ZU*) as a personal individual "continuum" of Awareness interacting with all Spheres of Existence on the Standard Model of Systemology; a spectrum of potential variations and interactions of a monistic continuum or singular *Spiritual Life Energy (ZU)* demonstrated on the Standard Model; an energetic channel of potential POV and "locations" of Beingness, demonstrated in early Systemology materials as an individual Alpha-Spirit's personal *Identity-continuum*, potentially connecting *Awareness (ZU)* of *Self* with "*Infinity*" simultaneous with all points considered in existence; a symbolic demonstration of the "*Life-line*" on which *Awareness (ZU)* extends from the direction of the "Spiritual Universe" (AN) in its true original *alpha state* through an entire possible range of activity resulting in its *beta state* and control of a *genetic-entity* occupying the *Physical Universe (KI)*.

**Zu-Vision** : the true and basic (*Alpha*) Point-of-View (perspective, POV) maintained by *Self* as *Alpha-Spirit* outside boundaries or considerations of the *Human Condition* "Mind-Systems" and *exterior* to beta-existence reality agreements with the Physical Universe; a POV of Self *as* "a unit of Spiritual Awareness" that exists independent of a "body" and entrapment in a *Human Condition*; "spirit vision" in its truest sense.

# Certificate of Completion

This certificate is awarded to

_____

for the successful completion of
the Mardukite Academy Basic Course on

## The Fundamentals of Systemology

# WOULD YOU LIKE TO KNOW MORE

? ? ?

# SYS♀EMOLOGY
## *The Pathway to Self-Honesty*

*A Basic Introduction to
Mardukite Systemology*

## THE WAY INTO
## THE FUTURE

### *A Handbook for
the New Human*

a collection of
writings by
Joshua Free
selected by
James Thomas

Here are the basic answers to what has held Humanity back
from achieving its ultimate goals and unlocking the true
power of the Spirit and highest state of Knowing and Being.

"*The Way Into The Future*" illuminates the *Pathway*
leading to Planet Earth's true "metahuman" destiny.
With <u>excerpts</u> *from* "*Tablets of Destiny,*" "*Crystal Clear,*"
"*Systemology—The Original Thesis*" and "*The Power of Zu.*"
You can help shine clear light on anyone's pathway!

Carefully selected by Mardukite Publications Officer,
James Thomas, this critical *collection of eighteen
articles, lecture transcripts and reference chapters* by
Joshua Free is sure to be not only a treasured part
of your personal library, but also the perfect
introduction for all friends, family and loved ones.

(*Basic Grade-III Introductory Pocket Anthology*)

# SYSTEMOLOGY
## *The Pathway to Self-Honesty*
### ORIGINAL UNDERGROUND INTRODUCTIONS
### REVISED AND REISSUED IN HARDCOVER

## SYSTEMOLOGY
### *The Original Thesis of Mardukite New Thuoght*
by Joshua Free
(*Mardukite Systemology Liber-S-1X*)

The very first underground discourses released to the "New Thought" division of the Mardukite Research Organization privately over a decade ago and providing the inspiration for rapid futurist spiritual technology called "Mardukite Systemology."

## THE POWER OF ZU
### *Applying Mardukite Zuism & Systemology to Everyday Life*
by Joshua Free
Foreword by Reed Penn
(*Mardukite Systemology Liber-S-1Z*)

A unique introductory course on Mardukite Zuism & Systemology, including transcripts from a 3-day lecture series given by Joshua Free in December 2019 to launch the Mardukite Academy of Systemology & Founding Church of Mardukite Zuism just in time for the 2020's.

# SYSTEMOLOGY
## *The Pathway to Self-Honesty*

## THE TABLETS OF DESTINY REVELATION

**How Long-Lost Anunnaki Wisdom Can Change the Fate of Humanity**

by Joshua Free

*Mardukite Systemology Liber-One*

*second edition*

Discover the origins of the Pathway to Self-Honesty with the book that started it all!

In this newly revised "Revelations" Academy Edition: Rediscover the original system of perfecting the Human Condition on a Pathway that leads to Infinity. Here is a way!—a map to chart spiritual potential and redefine the future of what is means to be human.

A landmark public debut for Grade-III Systemology and the foundation stone for reaching higher and taking back control of your DESTINY!

The revelation of 6,000 year old secrets, providing the tools and wisdom to unlock human potential...

# SYSTEMOLOGY
## The Pathway to Self-Honesty

**CRYSTAL CLEAR**
**Handbook for Seekers**

*Achieving*
*Self-Actualization*
*& Spiritual Ascension*
*in This Lifetime*

by Joshua Free

*Mardukite Systemology*
*Liber-2B*

*second edition*

Take control of your destiny and chart the first steps
toward your own spiritual evolution.

Realize new potentials of the Human Condition with
a Self-guiding handbook for Self-Processing
toward Self-Actualization in Self-Honesty using actual
techniques and training provided for the coveted
"Mardukite Self-Defragmentation Course Program"
—once only available directly and privately from the
underground International Systemology Society.

Discover the amazing power behind the
applied spiritual technology
used for counseling and advisement in
the Mardukite Zuism tradition.

# SYS𝗧EMOLOGY
## *The Pathway to Self-Honesty*

## METAHUMAN DESTINATIONS

*The Original 2020 Professional Piloting Academy Course for Grade IV*

by Joshua Free

*Mardukite Systemology Liber-Two (2C,2D,3C) Revised 2-Volume Set*

*available individually*

Drawing from the Arcane Tablets and nearly a year of additional research, experimentation and workshops since the introduction of applied spiritual technology and systematic processing methods, Joshua Free provides the ground-breaking manual for those seeking to correct—or "defragment"—the conditions that have trapped viewpoints of the Spirit into programming and encoding of the Human Condition.

Experience the revolutionary professional course in advanced spiritual technology for Mardukite Systemologists to "Pilot" the way to higher ideals that can free us from the Human Condition and return ultimate command and control of creation to the Spirit.

# SYSTEMOLOGY
## The Gateways to Infinity

## IMAGINOMICON

**Accessing the Gateway to Higher Universes**

**A New Grimoire for the Human Spirit**

by Joshua Free

*Mardukite Systemology Grade-IV Metahumanism, Wizard Level-0, Liber-3D*

*revised edition*

The Way Out. Hidden for 6,000 Years.
*But now we've found the Key.*
A grimore to summon and invoke, command and control,
the most powerful spirit to ever exist.
*Your Self.*

*Access* beyond physical existence.
*Fly* free across all Gateways.
Go back to where it all began and reclaim that
*personal universe* which the *Spirit* once called "*Home.*"

*Break free from the Matrix;*
control the Mind and command the Body
from outside those systems
— because *You* were never "human" —
fully realize what it means to be a *spiritual being,*
then rise up through the Gateways to Higher Universes
and *BE.*

# SYSTEMOLOGY

*The Gateway to Infinite Self-Honesty*

## THE WAY OF THE WIZARD

*Utilitarian Systemology*

**A New Metahuman Ethic**

by Joshua Free

*Mardukite Systemology Liber-3E*

*Grade-IV to Grade-V transition bridge*

Your ticket off of a Prison Planet...
...and a Pathway leading to Spiritual Ascension!

Accumulated involvement in dangerous situations, states of confusion, unjust destruction and being at the effect end of faulty—or—blatantly false information, all lend to fragmented purposes that may very well be painted to appear "for our own good." Instead they are non-survival or counter-survival oriented, leading us away from routes to achieve "greater heights"—higher, more ideal, states of knowingness and beingness—including the Magic Universe immediately preceding this one.

Here then is a bridge from Grade-IV to Grade-V, the next great frontier of the *Pathway* crossed by participants in the "Freedom From" workshops led by Joshua Free in 2021.

# SYSTEMOLOGY
## The Gateways to Infinity

## SYSTEMOLOGY-180
### The Fast-Track to Ascension

*A Handbook for Pilots*

by Joshua Free

*Mardukite Grade-V
Systemology
Liber-180*

*Expert application of
all Grade-III and Grade-IV
training and techniques*

A perfected "metahuman" state for the Human Condition awaits; free of emotional turbulence, societal programming and an ability to be truly Self-Determined from the clear perspective of the actual Self, the Eternal Spirit or "I-AM" Awareness that is back of and beyond this existence—an "Angel" or "god" that has fallen only by its own considerations, by being convinced that it resides locally here on earth within a perishable human shell.

"*Systemology-180*" presents newly revised instruction from the Mardukite Academy to deliver the fastest results in climbing the Ladder of Ascension. Hundreds of exercises and techniques that progressively free you from bonds of the Human Condition and increase your spiritual horsepower enough to break the chains and attachments to the material world and an existence confined to a material body.

# SYSⲐEMOLOGY
## *The Gateways to Infinity*

### SYSTEMOLOGY:
## BACKTRACK
### Reclaiming Spiritual Power & Past-Life Memory

by Joshua Free

*Mardukite Grade-V Systemology Liber-4*

*Transcripts of the original lectures with diagrams and glossary*

We are all Spiritual Beings that have known a very long existence. Even before the evolution of Humans or Earth, we existed as other forms, in other times and spaces. We have descended down a very long *track* of potential Beingness and considerations, a *track* that parallels the allegory of "Fallen Angels" enticed by mundane bodies; only to be trapped in them and longing to *Ascend* again.

*What if we could recover the long forgotten Knowingness of our past existences? What if we could reclaim our true Spiritual power that we have lost sight of? What if we could actually Backtrack our descent and return to the Source?*

"Backtrack" documents the first advanced course given by Joshua Free to the Systemology Society for Grade-V. He candidly introduces the new Wizard-Level subject of Alpha-Defragmentation to Grade-III and Grade-IV alumni ready to embark on their next phase of the *Pathway*.

*Commemorating the Mardukite 15th Anniversary!*

# NECRONOMICON
## THE COMPLETE ANUNNAKI BIBLE
(*Deluxe Edition Hardcover Anthology*)
*collected works by Joshua Free*

The ultimate masterpiece of Mesopotamian magic, spirituality and history, providing a complete collection—a grand symphony—of the most ancient writings on the planet. The oldest Sumerian and Babylonian records reveal detailed accounts of cosmic history in the Universe and on Earth, the development of human civilization and descriptions of world order. All of this information has been used, since ancient times, to maintain spiritual and physical control of humanity and its systems. It has proved to be the predecessor and foundation of all global scripture-based religious and mystical traditions thereafter. These are the raw materials, unearthed from the underground, which have shaped humanity's beliefs, traditions and existence for thousands of years—right from the heart of the Ancient Near East: Sumer, Babylon and even Egypt...

Also available abridged in hardcover as:
*"Necronomicon: The Compact Anunnaki Bible"*

∞

PUBLISHED BY THE **JOSHUA FREE** IMPRINT REPRESENTING

**The Founding Church of Mardukite Zuism
& Mardukite Academy of Systemology**

**mardukite.com**

Printed in the USA
CPSIA information can be obtained
at www.ICGtesting.com
LVHW021626061023
760217LV00030B/557/J

9 781961 509245